Cover art by V. Yankelevski

I.M. Gel'fand   E.G. Glagoleva   E.E. Shnol

Functions and Graphs

1990

Birkhäuser
Boston · Basel · Berlin

I. M. Gelfand
Department of Mathematics
Rutgers University
New Brunswick, NJ 08903

E.G. Glagoleva
University of Moscow
117234 Moscow
USSR

E.E. Shnol
University of Moscow
117234 Moscow
USSR

Functions and Graphs was originally published in 1966 in the Russian language under the title Funktsii i grafiki.

The english language edition was translated and adapted by Leslie Cohn and David Sookne under the auspices of the Survey of Recent East European Mathematical Literature, conducted by the University of Chicago under a grant from the National Science Foundation. It is republished here with permission from the University of Chicago for its content and the MIT Press for its form.

Library of Congress Cataloging-in-Publication Data
Gel'fand, I. M. (Izrail ' Moiseevich)
  [Funktsii i grafiki. English]
  Functions and graphs / I.M. Gelfand, E.G. Glagoleva, and E.E.
  Shnol ; translated and adopted from the second Russian edition by
  Thomas Walsh and Randell Magee.
    p. cm.
  Translation of Funktsii i grafiki.
  ISBN 0-8176-3532-7
  1. Functions. 2. Algebra--Graphic methods. I. Glagoleva, E. G.
  (Elena Georgievna) II. Shnol', E. E. (Emmanuil El'evich)
  III. Title.
  QA331.G3613 1990   90-48213
  515--dc20   CIP

Printed on acid-free paper.

ISBN 0-8176-3532-7   SPIN 10882894
ISBN 3-7643-3532-7

Text reproduced with permission of the MIT Press, Cambridge, Massachusetts, from their edition published in 1967.

Printed and bound by Quinn-Woodbine, Woodbine, NJ.

Printed in the U.S.A.

9 8 7 6

Preface

Dear Students,

We are going to publish a series of books for high school students. These books will cover the basics in mathematics. We will begin with algebra, geometry and calculus. In this series we will also include two books which were written 25 years ago for the Mathematical School by Correspondence in the Soviet Union. At that time I had organized this school and I continue to direct it.

These books were quite popular and hundreds of thousands of each were sold. Probably the reason for their success was that they were useful for independent study, having been intended to reach students who lived in remote places of the Soviet Union where there were often very few teachers in mathematics.

I would like to tell you a little bit about the Mathematical School by Correspondence. The Soviet Union, you realize, is a large country and there are simply not enough teachers throughout the country who can show all the students how wonderful, how simple and how beautiful the subject of mathematics is. The fact is that everywhere, in every country and in every part of a country there are students interested in mathematics. Realizing this, we organized the School by Correspondence so that students from 12 to 17 years of age from any place could study. Since the number of students we could take in had to be restricted to about 1000, we chose to enroll those who lived outside of such big cities as Moscow, Leningrad and Kiev and who inhabited small cities and villages in remote areas. The books were written for them. They, in turn, read them, did the problems and sent us their solutions. We never graded their work -- it was forbidden by our rules. If anyone was unable to solve a problem then some personal help was given so that the student could complete the work.

Of course, it was not our intention that all these students who studied from these books or even completed the School should choose mathematics as their future career. Nevertheless, no matter what they would later choose, the results of this training remained with them. For many, this had been their first experience in being able to do something on their own -- completely independently.

I would like to make one comment here. Some of my American colleagues have explained to me that American students are not really accustomed to thinking and working hard, and for this reason we must make the material as attractive as possible. Permit me to not completely agree with this opinion. From my long experience with young students all over the world I know that they are curious and inquisitive and I believe that if they have some clear material presented in a simple form, they will prefer this to all artificial means of attracting their attention -- much as one buys books for their content and not for their dazzling jacket designs that engage only for the moment.

The most important thing a student can get from the study of mathematics is the attainment of a higher intellectual level. In this light I would like to point out as an example the famous American physicist and teacher Richard Feynman who succeeded in writing both his popular books and scientific works in a simple and attractive manner.

I.M. Gel'fand

Foreword

This book is called *Functions and Graphs*. As I mentioned, this book was written 25 years ago. I believe that today we would write in the same way.

The process of drawing graphs is a way of transferring your formulas and your data into geometrical form. Thus, drawing graphs is one of the ways to "see" your formulas and your functions and to observe the way in which this function changes. For example, if it is written $y = x^2$ then you already see a parabola [see picture 1]; if $y = x^2 - 4$ then you see a parabola dropped by four [see picture 2]; and if $y = 4 - x^2$ then you see the previous graph turned upside down.

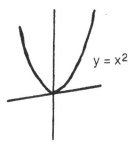

This skill, to see simultaneously the formula and its geometrical representation, is very important not only for studies in mathematics but in studies of other subjects as well. It will be a skill that will remain with you for the rest of your life, like riding a bicycle, typing, or driving a car. Graphs are widely used in economics, engineering, physics, biology, applied mathematics, and of course in business.

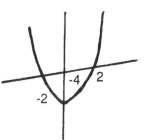

This book, along with the others in this series, is not intended for quick reading. Each section is designed to be studied carefully. You can first scan the section and choose what is interesting for you. Nor is it necessary to solve all of the problems. Choose what you like. And if it is difficult for you, come back to it and try to understand what made it hard for you.

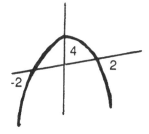

One more remark. This book as well as other books in this series is intended to be compatible with computers. However, do not think for one moment that in your study of mathematics you will be able to rely solely on computers. Computers can help you solve a problem. The computer cannnot -- nor will it ever be able -- to think and understand like you can. But of course it can sometimes inspire you to think and to understand.

Note to Teachers
This series of books includes the following material:

1. *Functions and Graphs*
2. *The Method of Coordinate*s
3. *Algebra*
4. *Geometry*
5. *Calculus*
6. *Combinatorics*

Of course, all of the books may be studied independently. We would be very grateful for your comments and suggestions. They are especially valuable because books 3 through 6 are in progress and we can incorporate your remarks. For the book *Functions and Graphs* we plan to write a second part in which we will consider other functions and their graphs, such as cubic polynomials, irrational functions, exponential function, trigonometrical functions and even logarithms and equations.

Contents

Preface v

Foreword vii

Introduction 1

Chapter 1   Examples 9

Chapter 2   The Linear Function 22

Chapter 3   The Function $y = |x|$ 26

Chapter 4   The Quadratic Trinomial 39

Chapter 5   The Linear Fractional Function 55

Chapter 6   Power Functions 66

Chapter 7   Rational Functions 80

Problems for Independent Solution 89

Answers and Hints to Problems and Exercises
Marked with the Sign $\oplus$ 103

—

Introduction

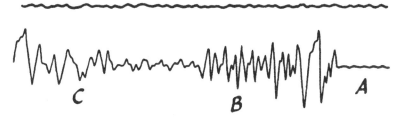

Fig. 1

In Fig. 1 the reader can see two curves traced by a seismograph, an instrument which registers fluctuations of the earth's crust. The upper curve was obtained while the earth's crust was undisturbed, the lower shows the signals of an earthquake.

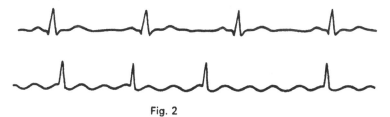

Fig. 2

In Fig. 2 there are two cardiograms. The upper shows normal heartbeat. The lower is taken from a heart patient.

Figure 3 shows the so-called characteristic of a semiconducting element, that is, the curve displaying the relationship between current intensity and voltage.

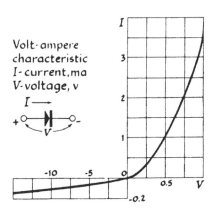

Fig. 3

In analyzing a seismogram, the seismologist finds out when and where an earthquake occurred and determines the intensity and character of the shocks. The physician examining a patient can, with the help of a cardiogram, judge the disturbances in heart activity; a study of the cardiogram helps to diagnose a disease correctly. The characteristic curve of a semiconducting element enables the radio-electrical engineer to choose the most favorable condition for his work.

All of these people study certain functions by the graphs of these functions.

What, then, is a function, and what is the graph of a function?

Before giving a precise definition of a function, let us talk a little about this concept. Descriptively speaking, a function is given when to each value of some quantity, which mathematicians call the argument and usually denote by the letter x, there corresponds the value of another quantity y, called the function.

Thus, for example, the magnitude of the displacement of the earth's surface during an earthquake has a definite value at each instant of time; that is, the amount of displacement is a function of time. The current intensity in a semiconducting element is a function of voltage, since to each value of the voltage there corresponds a definite value of current intensity.

Many such examples can be mentioned: the volume of a sphere is a function of its radius, the altitude to which a stone rises when thrown vertically upward is a function of its initial speed, and so on.

Let us now pass to the precise definitions. To say that the quantity y is a function of the quantity x, one first of all specifies which values x can take. These "allowed" values of the argument x are called *admissible values*, and the set of all admissible values of the quantity or variable x is called the *domain of definition* of the function y.

For instance, if we say that the volume V of the sphere is a function of its radius R, then the domain of definition of the function $V = \frac{4}{3}\pi R^3$ will be all numbers greater than zero, since the value R of the radius of the sphere can be only a positive number.

Whenever a function is given, it is necessary to specify its domain of definition.

Definition I

We say that y is a *function* of the variable x, if: (1) it is specified which values of x are admissible, i.e., the domain of definition of the function is given, and (2) to each admissible value of x there corresponds exactly one value of the variable y.

3

Instead of writing "the variable y is a function of the variable x", one writes

$$y = f(x).$$

(Read: "y is equal to f of x.")

The notation $f(a)$ denotes the numerical value of the function $f(x)$ when x is equal to a. For example, if

$$f(x) = \frac{1}{x^2 + 1},$$

then

$$f(2) = \frac{1}{2^2 + 1} = \frac{1}{5},$$

$$f(1) = \frac{1}{1^2 + 1} = \frac{1}{2},$$

$$f(0) = \frac{1}{0^2 + 1} = 1, \text{ etc.}$$

The rule by which for each value of x the corresponding value of y is found can be given in different ways, and no restrictions are imposed on the form in which this rule is expressed. If the reader is told that y is a function of x, then he has only to verify that: (1) the domain of definition is given, that is, the values that x can assume are specified, and (2) a rule is given whereby to each admissible value of x there can be associated a unique value of y.

What kind of rule can this be?

Let us mention some examples.

1. Suppose we are told that x may be any real number and y can be found by the formula

$$y = x^2.$$

The function $y = x^2$ is thus given by a formula.

The rule may also be verbal.

2. The function y is given in the following manner: If x is a positive number, then y is equal to 1; if x is

4

a negative number, then y is equal to -1; if x is equal to zero, then y is equal to zero.

Let us mention yet another example of a function given by a verbal rule.

3. Every number x can be written in the form $x = y + \alpha$, where α is a nonnegative number less than one, and y is an integer. It is clear that to each number x there corresponds a unique number y; that is, y is a function of x. The domain of definition of this function is the entire real axis. This function is called "the integral part of x" and is denoted thus:

$$y = [x].$$

For example,

$$[3.53] = 3, \quad [4] = 4, \quad [0.3] = 0, \quad [-0.3] = -1.$$

We shall use this function later in our exercises.

4. Let us consider the function $y = f(x)$, defined by the formula

$$y = \frac{\sqrt{x + 3}}{x - 5}.$$

What can reasonably be considered as its domain of definition?

If a function is given by a formula, then usually its so-called *natural* domain of definition is considered, that is, the set of all numbers for which it is possible to carry out the operations specified by the formula. This means that the domain of definition of our function does not contain the number 5 (since at $x = 5$ the denominator of the fraction vanishes) and values of x less than -3 (since for $x < -3$ the expression under the root sign is negative). Thus, the natural domain of definition of the function

$$y = \frac{\sqrt{x + 3}}{x - 5}$$

consists of all numbers satisfying the relations:

$$x \geq -3, \qquad x \neq 5.$$

A function can be represented geometrically with the help of a graph. In order to construct a graph of some function, let us consider some admissible value of x and the corresponding value of y. For example, suppose the value of x is the number a, and the corresponding value of y is the number $b = f(a)$. We represent this pair of numbers a and b by the point with the coordinates (a, b) in the plane. Let us construct such points for all admissible values of x. The collection of points obtained in this way is the graph of the function.

Definition II

The *graph of a function* is the set of points whose abscissas are admissible values of the argument x and whose ordinates are the corresponding values of the function y.

For example, Fig. 4 depicts the graph of the function

$$y = [x].$$

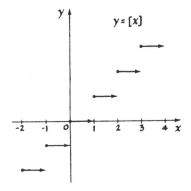

Fig. 4

It consists of an infinite set of horizontal line segments. The arrows indicate that the right end points of these segments do not belong to the graph (whereas the left end points belong to it and therefore are marked by a thick point).

A graph can serve as the rule by which the function is defined. For example, from the characteristic curve of a semiconducting element

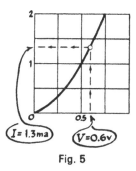

Fig. 5

it is possible to determine (see Fig. 5) that, if the argument V equals 0.6 volt, then the function I is equal to 1.3 milliamperes.

Representing functions by graphs is very convenient because, by looking at graphs, it is at once possible to distinguish one function from the other.

The reader is asked to look once more at the lower curve of Fig. 1. In this graph the most inexperienced person will at once recognize the signals of an earthquake (parts B and C). On closer examination, he will certainly also note the difference in character between the waves in parts B and C (a seismologist could point out that part B represents the so-called P-wave, the wave traveling deep in the earth's crust, while part C represents the S-wave traveling on the surface).

Try to distinguish these two sections by means of the tables of values arranged side by side. (We could not include here a table of values for the whole curve, because it would fill a whole page. In the margin the reader finds tables for small portions of sections B and C.)

Wave P (interval 0.2 sec)	Wave S (interval 0.4 sec)
0.1	0.2
0.1	0.5
−1.6	2.5
−1.7	4.9
−2.4	7.1
−3.0	6.1
−4.5	3.8
−3.8	0.4
−2.9	0.2
−1.1	0.7
0.8	1.5
3.3	2.5
5.1	3.2
3.7	2.8
0.0	0.4
−2.0	−2.2
−4.4	−3.3
−5.8	−4.5
−3.8	−4.8
−1.6	−4.8
	−4.8
	−3.7
	−3.5
	−4.4
	−6.6

7

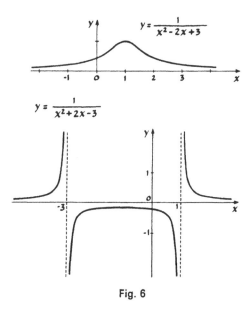

Fig. 6

Figure 6 shows graphs of two functions that are defined by very similar formulas:

$$y = \frac{1}{x^2 - 2x + 3} \quad \text{and} \quad y = \frac{1}{x^2 + 2x - 3}.$$

Of course, the difference in the behavior of these two functions can also be detected from their formulas. But if one looks at their graphs, this difference stands out immediately.

Whenever it is necessary to explain the general behavior of a function, to find its distinctive features, a graph is irreplaceable, by virtue of its visual character. Therefore, once an engineer or scientist has obtained a function in which he is interested, in the form of a formula or a chart, he usually takes a pencil and sketches a graph of this function, to find out how it behaves, what it "looks like."

CHAPTER 1

Examples

1

If the definition is followed literally, then in order to construct a graph of some function, it is necessary to find all pairs of corresponding values of argument and function and to construct all points with these co-ordinates. In the majority of cases it is practically impossible to do this, since there are infinitely many such points. Therefore, usually a few points belonging to the graph are joined by a smooth curve.

In this way, let us try to construct the graph of the function

$$y = \frac{1}{1 + x^2} . \qquad (1)$$

$$y = \frac{1}{1+x^2}$$

x	y
0	1
1	$\frac{1}{2}$
2	$\frac{1}{5}$
3	$\frac{1}{10}$

Table 1

Let us choose some values of the argument, find the corresponding values of the function, and write them down in a table (see Table 1). We construct the points with the computed coordinates and join them by a dotted line, for the time being (Fig. 1).

Let us now verify whether we have drawn the curve correctly between the points found to lie on the graph.

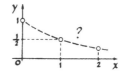

Fig. 1

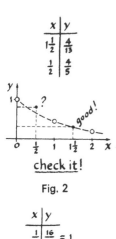

Fig. 2

check it!

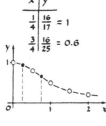

one
more
check

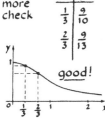

good!

Fig. 3

x	y
1	$\frac{1}{2}$
2	$\frac{1}{5}$
3	$\frac{1}{10}$
⋮	

x	y
-1	$\frac{1}{2}$
-2	$\frac{1}{5}$
-3	$\frac{1}{10}$
⋮	

Fig. 4

For this purpose let us take some intermediate value of the argument, say, $x = 1\frac{1}{2}$, and compute the corresponding value of the function $y = \frac{4}{13}$. The point obtained, $(1\frac{1}{2}, \frac{4}{13})$, falls nicely on our curve (Fig. 2), so that we have drawn it quite accurately.

Now we try $x = \frac{1}{2}$. Then $y = \frac{4}{5}$, and the corresponding point lies above the curve we have drawn (Fig. 2). This means that between $x = 0$ and $x = 1$ the graph does not go as we thought. Let us take two more values, $x = \frac{1}{4}$ and $x = \frac{3}{4}$, in this doubtful section. After connecting all these points, we get the more accurate curve represented in Fig. 3. The points $(\frac{1}{3}, \frac{9}{10})$ and $(\frac{2}{3}, \frac{9}{13})$, taken as a check, fit the curve nicely.

2

In order to construct the left half of the graph, it is necessary to fill in one more table for negative values of the argument. This is easy to do. For example,

for $x = 2$ we have $y = \dfrac{1}{2^2 + 1} = \dfrac{1}{5}$,

for $x = -2$ we have $y = \dfrac{1}{(-2)^2 + 1} = \dfrac{1}{5}$.

This means that together with the point $(2, \frac{1}{5})$, the graph also contains $(-2, \frac{1}{5})$, the point symmetric to the first with respect to the y-axis.

In general, if the point (a, b) lies on the right half of our graph, then its left half will contain the point $(-a, b)$ symmetric to (a, b) with respect to the y-axis (Fig. 4). Therefore, in order to obtain the left part of the graph of function (1) corresponding to negative values of x, it is necessary to reflect the right half of this graph in the y-axis.

Figure 5 shows the over-all form of the graph.

If we had been hasty and had used our original sketch for the construction of the part of the graph corresponding to negative x (Figs. 1 and 2), then it would have had a "kink" (corner) at $x = 0$. There is

10

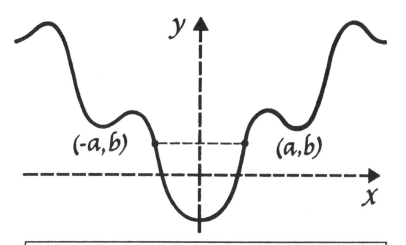

$$f(-a) = f(a)$$

If the values of some function corresponding to any two values of the argument equal in absolute value but with opposite signs (that is, the values a and -a) are equal, then such a function is called even. Any even function has a graph which is symmetric with regard to the y-axis.

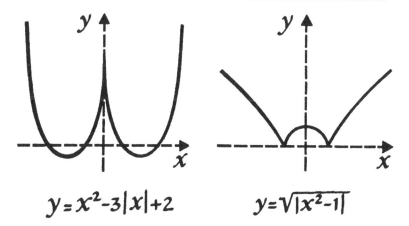

$y = x^2 - 3|x| + 2$

$y = \sqrt{|x^2 - 1|}$

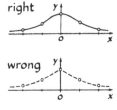
Fig. 5

no such kink in the accurate graph; instead there is a smooth "dome."

EXERCISES

1. The graph of the function

$$y = \frac{1}{3x^2 + 1} \qquad (2)$$

is similar to the graph of the function

$$y = \frac{1}{x^2 + 1}.$$

Construct it.

2. Which of the following functions are even (for the definition and the graph of an even function as well as some examples, see p. 11)?

(a) $y = 1 - x^2$;   (b) $y = x^2 + x$;

(c) $y = \dfrac{x^2}{1 + x^4}$;   (d) $y = \dfrac{1}{1 - x} + \dfrac{1}{1 + x}.$

3

Let us now take the function

$$y = \frac{1}{3x^2 - 1}. \qquad (3)$$

In form this formula differs little from Formula 2. However, in the construction of this graph by points, troubles immediately arise.

Let us again work out a table and plot the computed points in a diagram. It is not clear how to join these points; it seems that the point $(0, -1)$ does not "fit in" (Fig. 6).

Try to construct a graph of this function yourself. Do not become discouraged if you have to find more points than expected in order to understand how this curve behaves.

Thereafter you are urged to read, on pages 14 to 16 how the authors construct this graph and what useful conclusions can be drawn from the construction.

$y = \dfrac{1}{3x^2 - 1}$

x	y
0	1
1	$\frac{1}{2}$
2	$\frac{1}{11}$
-1	$\frac{1}{2}$
-2	$\frac{1}{11}$

Fig. 6

12

4

The graph of the polynomial

$$y = x^4 - 2x^3 - x^2 + 2x \qquad (4)$$

will also be constructed first by points.

Taking for the argument values equal to 0, 1, and 2, we obtain values of the function equal to zero. Let us now take $x = -1$. Again we obtain the result $y = 0$. The corresponding points of the graph $(0, 0)$, $(1, 0)$, $(2, 0)$, $(-1, 0)$ lie on the x-axis (Fig. 7).

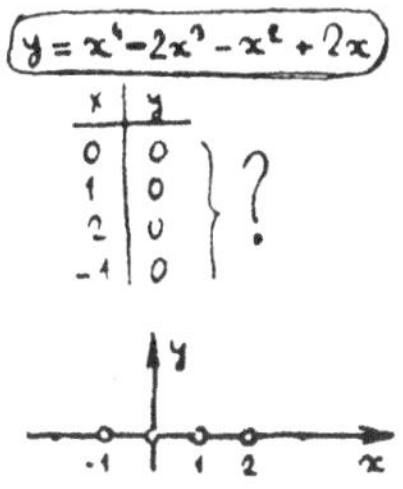

If we confine ourselves to these four values of the argument, then the x-axis will be a "smooth" curve joining the points obtained. It is clear, however, that the x-axis is not the graph of our function because the polynomial

$$x^4 - 2x^3 - x^2 + 2x$$

cannot equal zero for all values of x.

Let us take two more values of the argument, $x = -2$ and $x = 3$. The corresponding points $(-2, 24)$ and $(3, 24)$ do not lie on the x-axis; on the contrary, they are located very far from it (Fig. 8).

From the appearance of the graph, its shape is still unclear. Of course, it is possible to find a sufficient number of intermediate points and construct an approximation to the graph, as we did before, but this method is not very reliable.

We shall try to proceed differently.

Let us find out where the function is positive (and, therefore, the graph lies above the x-axis) and where it is negative (that is, the graph lies below the x-axis).

For this purpose, let us factor the polynomial that defines the function:

$$x^4 - 2x^3 - x^2 + 2x$$
$$= x^3(x - 2) - x(x - 2)$$
$$= (x^3 - x)(x - 2) = x(x^2 - 1)(x - 2)$$
$$= (x + 1)x(x - 1)(x - 2).$$

Fig. 7

Fig. 8

13

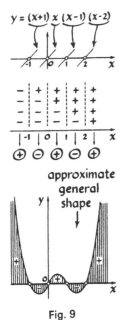

$y = (x+1) x (x-1) (x-2)$

−	+	+	+	+
−	−	+	+	+
−	−	−	+	+
−	−	−	−	+

⊕ ⊖ ⊕ ⊖ ⊕

approximate
general
shape
↓

Fig. 9

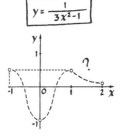

$y = \dfrac{1}{3x^2 - 1}$

Fig. 10

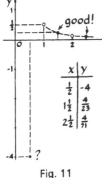

good!

x	y
$\frac{1}{2}$	-4
$1\frac{1}{2}$	$\frac{4}{23}$
$2\frac{1}{2}$	$\frac{4}{71}$

Fig. 11

It is now obvious that our function equals zero only at those four points which we have already plotted on the graph. To the left of the point $x = -1$, all four factors are negative, and the function is positive. Between the points $x = -1$ and $x = 0$ (that is, in the interval $-1 < x < 0$), the factor $x + 1$ becomes positive, while the remaining factors remain negative: the function is negative. In the region $0 < x < 1$, we have two positive and two negative factors: the function is positive. In the next interval the function is again negative. Finally, as the argument passes through the point $x = 2$, the last of the factors become positive: the function becomes positive.

The graph of the function now takes approximately the appearance given in Fig. 9.

5

We now pass to the construction of the graph of the function

$$y = \frac{1}{3x^2 - 1},$$

which we have already discussed on page 12.

Let us mark in the figure the points of the graph corresponding to the values $x = -1, 0, 1, 2$, and let us join them by a curve. As a result we obtain something like Fig. 10.

Let us now take $x = \frac{1}{2}$. We obtain $y = -4$, and the point $(\frac{1}{2}, -4)$ lies considerably below our curve. This means that between $x = 0$ and $x = 1$ the graph runs quite differently!

A more accurate course of the graph is represented in Fig. 11. If we take two more values, $x = 1\frac{1}{2}$ and $x = 2\frac{1}{2}$, the corresponding points fall quite nicely on our curve.

How, then, does the graph proceed between the points $x = 0$ and $x = 1$?

Let us take $x = \frac{1}{4}$ and $x = \frac{3}{4}$. We obtain $y = -\frac{16}{13} \approx -1\frac{1}{4}$ and $y = \frac{16}{11} \approx 1\frac{1}{2}$, respectively. The course of the graph between the points $x = 0$ and

14

$x = \frac{1}{2}$ thus has become somewhat clearer (Fig. 12), but the behavior of the function between $x = \frac{1}{2}$ and $x = \frac{3}{4}$ remains, as previously, obscure.

If we take a few more intermediate values between $x = \frac{1}{2}$ and $x = \frac{3}{4}$, we see that the corresponding points of the graph lie not on one but on two smooth curves, and the graph assumes approximately the appearance of Fig. 13.

The reader can now well understand that the pointwise construction of a graph is risky and lengthy. If we take few points, then it may turn out that we obtain an altogether false picture of the function. If, on the other hand, we take more points, there will be much superfluous work, and some doubt still remains as to whether we did not omit something significant. How are we to proceed?

Let us recall that when we constructed the graph $y = 1/(x^2 + 1)$ in the interval $2 < x < 3$ and $1 < x < 2$ no additional points were required, while in the interval $0 < x < 1$, it was necessary to find 5 more points. Similarly in the construction of the graph of $y = 1/(3x^2 - 1)$ the most work was required for the interval $0 < x < 1$, where the curve splits into two branches.

Is it not possible to isolate such "dangerous" regions beforehand?

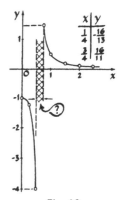

Fig. 12

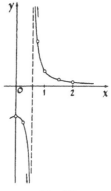

Fig. 13

6

Let us return for the third time to the graph of

$$y = \frac{1}{3x^2 - 1}.$$

If one looks at the expression defining the function, it is immediately obvious that for two values of x the denominator of this expression vanishes. These values are equal to $+\sqrt{\frac{1}{3}}$ and $-\sqrt{\frac{1}{3}}$, that is, approximately ± 0.58. One of them lies in the interval $\frac{1}{2} < x < \frac{3}{4}$, precisely where the function displays unusual behavior, where the graph does not go smoothly. It is now clear why this is the case.

15

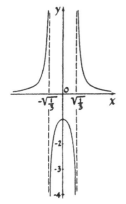

Fig. 14

In fact, for the values $x = \pm\sqrt{\tfrac{1}{3}}$, the function is not defined (division by zero is impossible); therefore, there can be no points on the graph with these abscissas — the graph does not intersect the straight lines $x = \sqrt{\tfrac{1}{3}}$ and $x = -\sqrt{\tfrac{1}{3}}$. Therefore, the graph decomposes into three separate branches. If x approaches one of the "forbidden" values, say, $x = \sqrt{\tfrac{1}{3}}$, then the fraction $1/(3x^2 - 1)$ increases without bond in absolute value: two branches of the graph approach the vertical straight line $x = \sqrt{\tfrac{1}{3}}$.

Our (even) function behaves analogously near the point $x = -\sqrt{\tfrac{1}{3}}$.

The general form of the graph of $y = 1/(3x^2 - 1)$ is shown in Fig. 14.

We now understand that *whenever a function is defined by a formula in the form of a fraction, it is necessary to focus attention on those values of the argument at which the denominator vanishes.*

7

What lesson, then, can be learned from the examples we have considered? In the study of the behavior of a function and in the construction of its graph, not all values of the argument are of equal importance. In the case of the function

$$y = \frac{1}{3x^2 - 1}$$

we saw the importance of those "special" points at which the function is not defined. The character of the graph of

$$y = x^4 - 2x^3 - x^2 + 2x$$

became clear to us when we found the points of intersection of the graph with the x-axis, namely, the roots of the polynomial.

In most cases the main part of work in the construction of graphs consists precisely in finding values of the argument significant for the given function and in investigating its behavior near these values. To com-

16

plete the construction of the graph after such an analysis, it usually suffices to find some intermediate values of the function between these characteristic points.

EXERCISES

1. Construct the graph of the function

$$y = \frac{1}{3x - 1} \, .$$

At what points does the graph intersect the co-ordinate axes?

Imagine that we placed the origin in the very center of a page in an exercise book and took 1 centimeter as the unit of measure (for definiteness we shall consider a page to be a rectangle of size 16 cm × 20 cm). Find the coordinates of the points at which the graph leaves the page.

2. Construct the graphs of the polynomials*

(a) $y = x^3 - x^2 - 2x + 2$;
(b) $y = x^3 - 2x^2 + x$. $\oplus$

(Notice that in case b the factorization of the polynomial results in two identical factors.)

8

Having constructed the graph of some function, we can use various methods to construct the graphs of some "related" functions.†

One of the simplest of these methods is the so-called *stretching along the y-axis*.

---

*The sign $\oplus$ marks problems and exercises to which there are answers at the end of the book (see p. 103).

†A method of this kind was already encountered on page 10, when we constructed the graph of $y = 1/(x^2 + 1)$. Having constructed the graph of this function for positive values of x, we were immediately able to construct the graph for negative x as well.

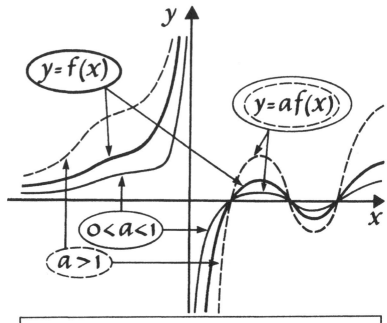

$$f(x) \longrightarrow af(x)$$

The graph of the function $y = af(x)$ is obtained from the graph of the function $y = f(x)$ by stretching it in the ratio $a:1$ along the y-axis (in case $|a| < 1$ we obtain contraction).

$$y = |x-1| + \tfrac{1}{2}(x-1) - \tfrac{1}{2}|x+1|$$

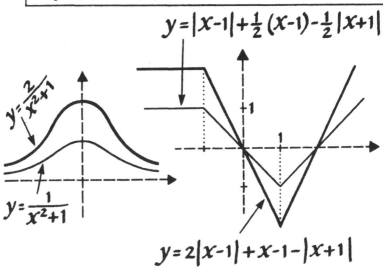

$$y = 2|x-1| + x - 1 - |x+1|$$

We constructed the graph of the function

$$y = \frac{1}{x^2 + 1}$$

(Fig. 5 on p. 12).

Let us now construct the graph of

$$y = \frac{3}{x^2 + 1}.$$

Let us take any point of the first graph, for example, $x = \frac{1}{2}$, $y = \frac{4}{5}$, that is, the point $M_1(\frac{1}{2}, \frac{4}{5})$. Clearly, we can obtain a point of the second graph by keeping x constant (that is, $x = \frac{1}{2}$) and increasing y in the ratio $3:1$. We thus obtain the point $M_2(\frac{1}{2}, \frac{12}{5})$. It can be obtained directly in the diagram (Fig. 15). For this it is necessary to increase the ordinate of the point $M(\frac{1}{2}, \frac{4}{5})$ in the ratio $3:1$.

If we carry out such a transformation with each point of the graph of $y = 1/(x^2 + 1)$, then the point $M(a, b)$ is transformed into the point $M'(a, 3b)$ of the graph of $y = 3/(x^2 + 1)$, and the entire graph, stretched in the ratio $3:1$ along the y-axis, turns into the graph of the function $y = 3/(x^2 + 1)$ (Fig. 16).

Thus, the graph of $y = 3/(x^2 + 1)$ represents the graph of $y = 1/(x^2 + 1)$ stretched in the ratio $3:1$ along the y-axis.

9

It is still easier to obtain the graph of the function

$$y = -\frac{1}{x^2 + 1}$$

from the graph of $y = 1/(x^2 + 1)$. In order to obtain a table for the function $y = -1/(x^2 + 1)$ from Table 1 on page 9 for the function $y = 1/(x^2 + 1)$, it is necessary only to change the sign of each of the numbers in the second column.

Then each point of the graph of $y = 1/(x^2 + 1)$, for example the point M with abscissa 2 and ordinate

$y = \dfrac{1}{x^2+1}$ $y = \dfrac{3}{x^2+1}$

x	y
0	1
$\frac{1}{2}$	$\frac{4}{5}$
2	$\frac{1}{5}$

x	y
0	3
$\frac{1}{2}$	$\frac{12}{5}$
2	$\frac{3}{5}$

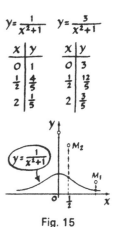

Fig. 15

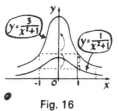

Fig. 16

19

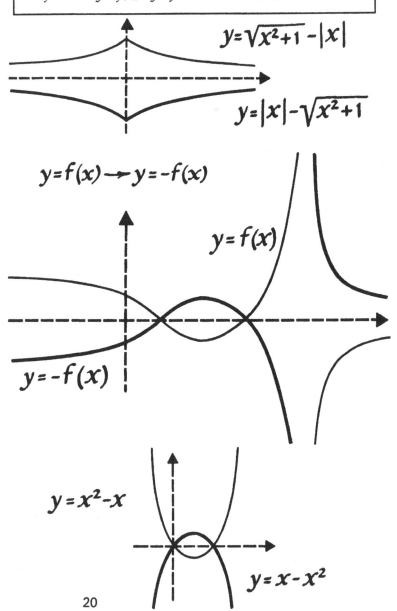

$$f(x) \longrightarrow -f(x)$$

The graph of y = -f(x) can be obtained from the graph of the function y = f(x) by reflection in the x-axis.

$y=\sqrt{x^2+1}-|x|$

$y=|x|-\sqrt{x^2+1}$

$y=f(x)\longrightarrow y=-f(x)$

$y=f(x)$

$y=-f(x)$

$y=x^2-x$

$y=x-x^2$

$\frac{1}{5}$, yields a point M' of the graph $y = -1/(x^2 + 1)$ with the same abscissa 2 but whose ordinate has the opposite sign $(-\frac{1}{5})$. Obviously the point $M'(2, -\frac{1}{5})$ is symmetric to the point $M(2, \frac{1}{5})$ with respect to the x-axis. Generally speaking, to the point $N(a, b)$ of the graph of $y = 1/(x^2 + 1)$ there corresponds the point $N'(a, -b)$ of the graph of $y = -1/(x^2 + 1)$.

Thus, the graph of the function $y = -1/(x^2 + 1)$ can be obtained from the graph of $y = 1/(x^2 + 1)$, by finding its mirror image in the x-axis (Fig. 17).

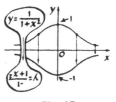

Fig. 17

EXERCISES

1. From the graph of $y = x^4 - 2x^3 - x^2 + 2x$ (Fig. 9, page 14), construct the graphs of $y = 3x^4 - 6x^3 - 3x^2 + 6x$ and $y = -x^4 + 2x^3 + x^2 - 2x$.

2. Construct the graph of $y = 1/(2x^2 + 2)$, using the graph of $y = 1/(x^2 + 1)$.

3. Construct the graphs of*

(a) $y = \frac{1}{2}[x]$;

(b) $y = x - [x]$ and $y = -2(x - [x])$;

(c) $y = [2x]$.

*The meaning of the symbol $[x]$, the integral part of the number x, was explained on page 5.

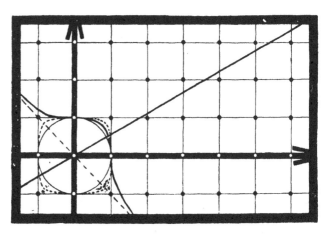

CHAPTER 2

The Linear Function

1

Let us now begin to study systematically the behavior of different functions and to construct their graphs. The characteristic behavior of functions and peculiarities of their graphs will be studied using the simplest examples. When we construct more complicated graphs, we shall try to find familiar elements in them.

The simplest function is the function $y = x$. The graph of this function is a straight line, the bisector of the first and third quadrants (Fig. 1).

In general, as you know, the graph of any linear function

$$y = kx + b$$

is some straight line. Conversely, any straight line not parallel to the y-axis is the graph of some linear function. The position of a straight line is completely determined by two of its points. Accordingly, a linear function is completely determined by giving its values for two values of the argument.

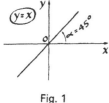

Fig. 1

22

1. Find the linear function

$$y = kx + b$$

that takes the value $y = 41$ at $x = -10$, and the value $y = 9$ at $x = 6$.

2. A straight line passes through the points $A(0, 0)$ and $B(a, c)$. Find the linear function whose graph is this straight line.

3. Draw a straight line through the origin at an angle of 60° with the y-axis. What function has this straight line as its graph?

4. (a) In Table 1 of the values of some linear function, two out of its five values are written down incorrectly. Find and correct them.

(b) The same question for Table 2.

5. Find the function

$$y = ax + b$$

if its graph is parallel to the graph of $y = x$ and passes through the point $(3, -5)$.

6. Find the linear function whose graph makes an angle of 60° with the x-axis and passes through the point $(3, -5)$.

7. The slope of a straight line* equals k. The straight line passes through the point $(3, -5)$. Find the linear function whose graph is this straight line.

2

A characteristic property of a linear function is that if x is increased uniformly, that is, by the same number for all x, then y also changes uniformly. Let us take the function $y = 3x - 2$. Suppose x takes the values

$$1, 3, 5, 7, \cdots,$$

each of which is larger than the preceding one, always

---

*The coefficient a is called the slope of the straight line $y = ax + b$.

Table 1

x	y
$\vdots$	$\vdots$
-2	-2
-1	3
0	1
1	2
2	-3

Table 2

x	y
-15	-33
-10	-13
0	7
10	17
15	27

by the same number, 2. The corresponding values of y will be

$$1, 7, 13, 19, \cdots .$$

The reader can see that each of the values of y is larger than the preceding one, also always by the same number, 6.

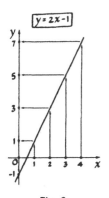

$y = 2x - 1$

Fig. 2

A sequence of numbers that is obtained from some number by constantly adding the same number forms an *arithmetic progression*. Thus, the characteristic property which we mentioned can be expressed in the following fashion: A linear function converts one arithmetic progression into another arithmetic progression. In our example (page 23), the function $y = 3x - 2$ converts the arithmetic progression 1, 3, 5, 7, ... into the arithmetic progression 1, 7, 13, 19, ... Figure 2 illustrates another example of how the function $y = 2x - 1$ changes the arithmetic progression 0, 1, 2, 3, ... into the arithmetic progression $-1, 1, 3, 5, 7, \ldots$.

EXERCISES

1. Find the linear function which would convert the arithmetic progression $-3, -1, 1, 3, \cdots$ into the arithmetic progression $-2, -12, -22, \cdots$, etc.

What linear function transforms the second progression into the first?

2. Suppose we are given two arithmetic progressions:

$$a, a + h, a + 2h, \cdots \text{ and } c, c + l, c + 2l, \cdots .$$

Is it always possible to find a linear function $y = kx + b$ which transforms the first progression into the second?

3. (a) The straight line $y = \frac{7}{15}x + \frac{1}{3}$ passes through two points with integral coordinates: $A(10, 5)$ and $B(-20, -9)$. Are there other "integral points" (points with integral coordinates) on this straight line?

24

(*b*) It is known that the straight line $y = kx + b$ passes through two integral points. Are there other integral points on this straight line?

(*c*) It is easy to construct a straight line that does not pass through any integral point. For example, $y = x + \frac{1}{2}$.

Is it possible for some straight line $y = kx + b$ to pass through just one integral point?*

*If you do not find an answer, look at Problem 4 on page 91.

CHAPTER 3

The Function $y = |x|$

1

Let us now consider the function

$$y = |x|,$$

where $|x|$ means the absolute value* or modulus of the number x.

Let us construct its graph, using the definition of absolute value. For positive x we have $|x| = x$; that is, this graph coincides with the graph of $y = x$ and is a half-line leaving the origin at an angle of 45° with the x-axis (Fig. 1). For $x < 0$, we have $|x| = -x$, which means that for negative x the graph of $y = |x|$ coincides with the bisector of the second quadrant (Fig. 2).

But then, the second half of the graph (for negative values of x) can easily be obtained from the first, if it

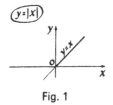

Fig. 1

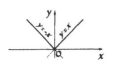

Fig. 2

*Let us recall: The absolute value of a positive number is equal to this number (if $x > 0$, then $|x| = x$); the absolute value of a negative number is equal to the number provided with the positive sign (if $x < 0$, then $|x| = -x$); the absolute value of zero is equal to zero ($|0| = 0$).

26

is noted that the function $y = |x|$ is even, since $|-a| = |a|$ (see the definition of an even function on page 11).

This means the graph of this function is symmetric with respect to the y-axis, and the second half of its graph can be obtained by reflection in the y-axis of the part traced out for positive x. This yields the graph represented in Fig. 3.

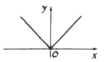

Fig. 3

2

Let us now construct the graph of

$$y = |x| + 1.$$

This graph can easily be constructed at once. We shall obtain it, however, from the graph of the function $y = |x|$. Let us work out a table of values of the function $y = |x| + 1$ and compare it with the same table worked out for $y = |x|$ by writing these tables side by side (Tables a, b). It is obvious that from each point of the first graph, $y = |x|$, a point of the second graph, $y = |x| + 1$, can be obtained by increasing y by 1. (For instance, the point $(-2, 2)$ of the graph of $y = |x|$ goes into the point $(-2, 3)$ of the graph of $y = |x| + 1$, located one above the first — see Fig. 4.) Hence, in order to get the points of the second graph, it is necessary to move each point of the graph up by 1; that is, the entire second graph is obtained from the first by an upward translation of 1 unit (see Fig. 4).

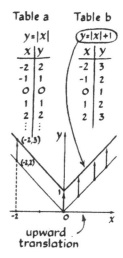

Table a

$y = |x|$

x	y
-2	2
-1	1
0	0
1	1
2	2

Table b

$y = |x| + 1$

x	y
-2	3
-1	2
0	1
1	2
2	3

upward translation

Fig. 4

Problem. Construct the graph of the function

$$y = |x| - 1.$$

Solution. Let us compare this graph with the graph of $y = |x|$. If the point $x = a, y = |a|$ lies on the first graph, then the point $x = a, y = |a| - 1$ will lie on the second graph. Therefore each point $(a, |a| - 1)$ of the second graph can be obtained from the point $(a, |a|)$ of the first graph by a downward translation of 1 unit, and the whole graph is obtained if the graph $y = |x|$ is moved downward 1 unit (Fig. 5).

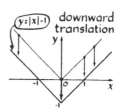

Fig. 5

27

Such a translation along the y-axis is useful in the construction of many graphs (see p. 29).

Suppose we are to construct the graph of the function

$$y = \frac{x^2 + 2}{x^2 + 1}.$$

Let us represent this function in the form

$$y = \frac{x^2 + 1 + 1}{x^2 + 1},$$

or

$$y = 1 + \frac{1}{x^2 + 1}.$$

It is now obvious that its graph can be obtained from the graph (Fig. 5 on p. 12) of $y = 1/(x^2 + 1)$ by an upward translation of 1 unit along the y-axis.

3

Let us now take the function

$$y = |x + 1|.$$

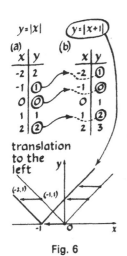

Fig. 6

The graph of this function will also be obtained from the graph of $y = |x|$. Let us again write two tables side by side, one for $y = |x|$ and one for $y = |x + 1|$ (Tables a, b). If we compare the values of the functions for the same x, then it turns out that for some x the ordinate of the first graph is larger than that of the second, and for some it is the other way around.

However, if we look closely at the right columns of these two tables, a connection between the tables can be established: The second function assumes the same values as the first, but does so one unit earlier, for smaller values of x. (Why?) Hence each point of the first graph, $y = |x|$, yields a point of the second graph, $y = |x + 1|$, moved one unit to the left; for example, the point $(-1, 1)$ gives rise to the point with

28

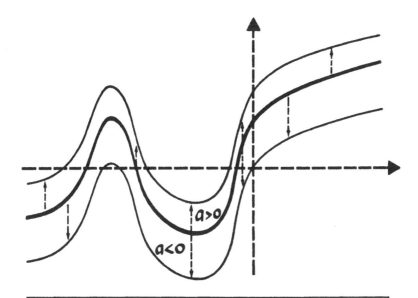

$$f(x) \longrightarrow f(x) + a$$

The graph of the function y = f(x) + a is obtained from the graph of the function y = f(x) by a translation along the y-axis of a units. The direction of the translation is determined by the sign of the number a (if a > 0, the graph moves up; if a < 0, the graph moves down).

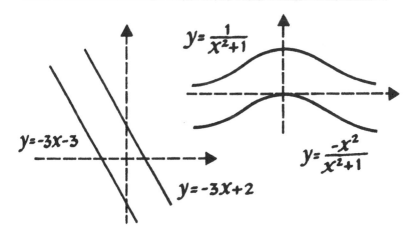

the coordinates $(-2, 1)$ (Fig. 6). Therefore, the whole graph of $y = |x + 1|$ is obtained if the graph of $y = |x|$ is moved to the left by one unit along the x-axis.

Problem. Construct the graph of the function

$$y = |x - 1|.$$

Solution. Let us compare it with the graph of $y = |x|$. If A is a point of the graph of $y = |x|$, with coordinates $(a, |a|)$, then the point $A'(a + 1, |a|)$ will be a point of the second graph with the same value for the ordinate y. (Why?) This point of the second graph can be obtained from the point $A(a, |a|)$ of the first graph by a translation to the right along the x-axis. Hence, the whole graph of $y = |x - 1|$ is obtained from the graph of $y = |x|$ by moving it to the right along the x-axis by 1 unit (Fig. 7). We can say the function $y = |x - 1|$ assumes the same values as the function $y = |x|$, but with a certain delay (a delay of 1 unit).

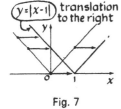

Fig. 7

Such a shift along the x-axis is useful in the construction of many graphs.

EXERCISES

1. Construct a graph of the function

$$y = \frac{1}{x^2 - 2x + 2}.$$

Hint: Represent the denominator of the fraction $1/(x^2 - 2x + 2)$ in the form $(x - 1)^2 + 1$.

2. State rules according to which from the graph of the function $y = f(x)$ it is possible to obtain the graphs of the functions $y = f(x + 5)$ and $y = f(x - 3)$.

3. Construct the graphs of $y = |x| + 3$ and $y = |x + 3|$.

30

4. Find all linear functions that at $x = 3$ assume the value $y = -5$.

Solution. Geometrically the condition can be formulated thus: find all straight lines passing through the point $(3, -5)$. Any (nonvertical) straight line passing through the origin is the graph of some function $y = kx$. Let us translate this straight line so that it passes through the required point $(3, -5)$, that is, 3 units to the right and 5 units downward (Fig. 8). After the first translation we obtain the equation

$$y = k(x - 3),$$

after the second

$$y = k(x - 3) - 5.$$

Answer. All linear functions, which at $x = 3$ assume the value $y = -5$, can be expressed by the formula

$$y = k(x - 3) - 5,$$

where k is any real number. (Compare this problem with Problem 7 on p. 23.)

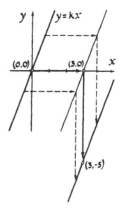

Fig. 8

Problem. Construct the graph of

$$y = |x + 1| + |x - 1|.$$

Solution. Let us first construct in one diagram the graphs of each of the terms:

$$y = |x + 1| \text{ and } y = |x - 1|.$$

The ordinate y of the desired graph is obtained by adding the ordinates of the two constructed graphs at the same point. Thus, for example, if $x = 3$, then the ordinate y_1 of the first graph is equal to 4, the ordinate y_2 of the second graph is equal to 2, and the ordinate y of the graph of $y = |x + 1| + |x - 1|$ is equal to 6.

31

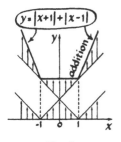

Fig. 9

Let us try to obtain the desired graph by adding at each point (that is, for each x) the ordinates of the two graphs. We thus obtain the drawing given in Fig. 9.

We see that the graph of $y = |x + 1| + |x - 1|$ is a broken line, composed of portions of three straight lines. Hence, in each of three intervals, the function varies linearly.

EXERCISES

1. Write down equations for each part of the broken line
$$y = |x + 1| + |x - 1|.$$

(**Answer.** for $x \leq -1$, $y = \cdots x + \cdots$;
         for $-1 \leq x \leq 1$, $y = \cdots$;
         for $x \geq 1$, $y = \cdots .$)

2. Where are the breaks in the broken line which is the graph of the function $y = |x| + |x + 1| + |x + 2|$? Find the equations of each of its parts.

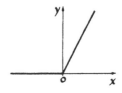

Fig. 10

3. (*a*) The function whose graph is represented in Fig. 10 can be defined by the following conditions:

$$\text{for } x < 0, \quad y = 0$$
$$\text{for } x \geq 0, \quad y = 2x.$$

Try to give this function in terms of one formula.

(*b*) Write down formulas for the functions whose graphs are represented in Figs. 11 and 12, respectively. $\oplus$

32

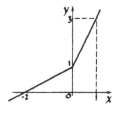

Fig. 11 Fig. 12

4. Construct the graph of the function

$$y = |3x - 2|.$$

Hint: Obtain this graph from the graph of $y = |x|$ by two transformations: a translation along the x-axis and stretching along the y-axis. In order to determine the correct value of the translation, it is necessary to pull the coefficient of x in front of the absolute value sign: $|3x - 2| = 3|x - \frac{2}{3}|$.

5

Problem. Construct the graph of

$$y = |2x - 1|.$$

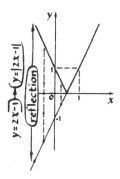

Fig. 13

Solution. We shall obtain this graph from the straight line $y = 2x - 1$ (Fig. 13). Wherever the straight line is above the x-axis, y is positive; that is, $2x - 1 > 0$. Hence, in this interval $|2x - 1| = 2x - 1$, and the desired graph coincides with the graph of $y = 2x - 1$. Wherever $2x - 1 < 0$ (that is, the straight line $y = 2x - 1$ is below the x-axis), $|2x - 1| = -(2x - 1)$. Thus, in order to obtain the graph of $y = |2x - 1|$ from this section of the graph of $y = 2x - 1$, it is necessary to change the sign of the ordinate of each point of the straight line $y = 2x - 1$, that is, to reflect this straight line in the x-axis. We thus obtain Fig. 14.

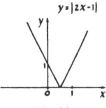

Fig. 14

33

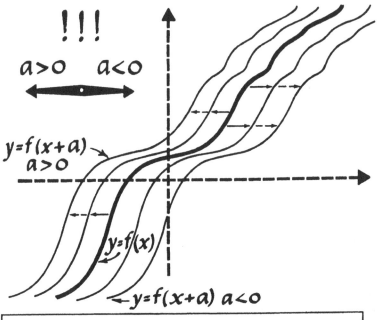

!!!

a>0   a<0

y=f(x+a)
  a>0

y=f(x)

y=f(x+a) a<0

$$f(x) \longrightarrow f(x+a)$$

The graph of the function y = f(x+a) is obtained from the graph of the function y = f(x) by a translation along the x-axis by -a units. The minus sign means that, if a is positive, the graph moves to the left, and if a is negative, the graph moves to the right.

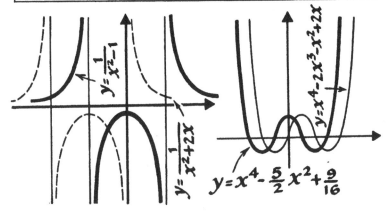

$y = \dfrac{1}{x^2-1}$

$y = \dfrac{1}{x^2+2x}$

$y = x^4 - 2x^3 - x^2 + 2x$

$y = x^4 - \dfrac{5}{2}x^2 + \dfrac{9}{16}$

EXERCISE

From the known graph of

$$y = x^4 - 2x^3 - x^2 + 2x$$

(Fig. 9, p. 14), construct the graph of

$$y = |x^4 - 2x^3 - x^2 + 2x|.$$

6

Problem. From the known graph of

$$y = \frac{1}{x^2 - 2x + 2} \qquad (1)$$

(Fig. 15), construct the graph of

$$y = \frac{1}{x^2 - 2|x| + 2}. \qquad (2)$$

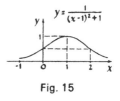

Fig. 15

Solution. Since for positive values of the argument $|x| = x$,

$$\frac{1}{x^2 - 2|x| + 2} = \frac{1}{x^2 - 2x + 2}.$$

Hence, to the right of zero the graph of Eq. 2 coincides with the graph of Eq. 1 (Fig. 16). In order to obtain the left half of the desired graph of Eq. 2, we note that the function $y = 1/(x^2 - 2|x| + 2)$ is even. This means that the left half of the graph of Eq. 2 is obtained from its right half by reflection in the y-axis (Fig. 17). The same is true in the general case: in order to obtain the graph of $y = f(|x|)$ from the graph of $y = f(x)$, it is necessary to reflect the half of the first graph lying to the right of the y-axis in the y-axis.

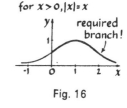

Fig. 16

EXERCISES

1. Construct the graph of $y = 2|x| - 1$.
2. Construct the graphs of

(*a*) $y = 4 - 2x$;

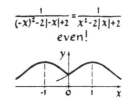

Fig. 17

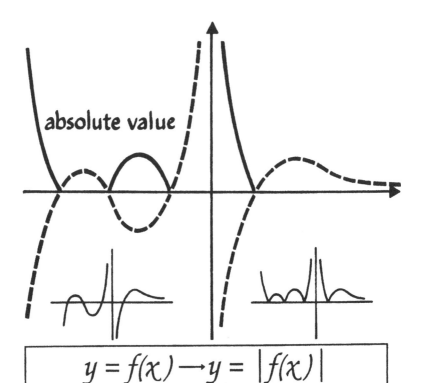

absolute value

$$y = f(x) \longrightarrow y = |f(x)|$$

In order to obtain the graph of $y = |f(x)|$ from the graph of $y = f(x)$, it is necessary to leave the parts of the graph lying above the x-axis unchanged, and to reflect the parts lying below the x-axis in the x-axis.

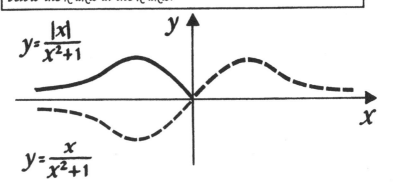

$y = \dfrac{|x|}{x^2+1}$

$y = \dfrac{x}{x^2+1}$

(b) $y = |4 - 2x|$;

(c) $y = 4 - 2|x|$;

(d) $y = |4 - 2|x||$.

3. Find the least value of the function

$$y = |x - 2| + |x| + |x + 2| + |x + 4|. \quad \oplus$$

In concluding this section we suggest that you solve some problems. At first glance they seem to bear no relation whatever to what we have been concerned with in this section, but on reflection you will see that this is not so.

Problems*

1. Seven matchboxes are arranged in a row. The first contains 19 matches, the second 9, and the following ones contain 26, 8, 18, 11, and 14 matches, respectively (Fig. 18). Matches may be taken from any box and put into any adjacent box. The matches must be redistributed so that their number in all boxes becomes the same. How can this be accomplished, shifting as few matches as possible?

Fig. 18

Solution. There is a total of 105 matches in all the boxes. Hence, if there were the same number of matches in all boxes, then each box would contain 15 matches. With such an arrangement of the boxes the problem has but one solution: Shift 4 matches from the first box to the second. As a result the first box will contain 15 matches and the second 13 matches. Take 2 matches from the third box and place them in the second, so that 24 matches remain in the third box. Shift the excess number of matches from this box into the fourth, and so on.

Problems 2 and 3 are somewhat more difficult. The question will be the same as in Problem 1.

*Problems 1 through 4 and the method of their solution were suggested by M. L. Tsetlin.

37

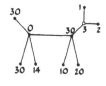

Fig. 19

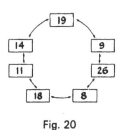

Fig. 20

2. Seven matchboxes are arranged in a straight line, as previously, but the number of matches in the boxes is different. That is, the first contains 1 match, the second contains 2, the following ones 3, 72, 32, 20, and 10.

3. The matchboxes are arranged in the shape of a "dog" (Fig. 19). Matches may be moved only along the lines (routes) joining the boxes.

Graphs can be used to advantage for the following problem.

4. Seven matchboxes are arranged in a circle. The first contains 19 matches, the second 9 matches, and the remaining ones contain 26, 8, 18, 11, and 14 matches, respectively (Fig. 20). It is permissible to move the matches from any box to any adjacent box. The matches must be shifted in such a way that the number of matches in all boxes becomes the same. How can this be done, shifting as few matches as possible? ⊕

38

CHAPTER 4

The Quadratic Trinomial

1

Let us now pass to the function

$$y = x^2.$$

You have, of course, constructed its graph, and you know that this curve has the special name of *parabola*. Graphs of the functions $y = ax^2$ are obtained from the graph of $y = x^2$ by stretching and are also called parabolas.*

EXERCISE

Figure 1 represents a parabola. (*a*) It is known that it is the graph of the function $y = x^2$. Find the scale of the diagram (the scale is the same for both axes). (*b*) What scale unit must be taken along the axes in order that the same curve serve as graph of the function $y = 5x^2$?

*It is interesting that all parabolas are similar to one another (see p. 43 as well as p. 95, Problem 16*d*).

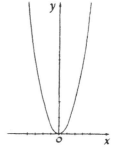

Fig. 1

39

2

Let us find out how the values of the function $y = x^2$ will change if the values of the argument constantly change by the same quantity; that is, if they form an arithmetic progression. For simplicity let us consider positive values of x. For example, suppose x takes the values

$$1, 2, 3, 4, \cdots, \text{etc.}$$

Then y will assume the values

$$1, 4, 9, 16, \cdots, \text{etc.}$$

You can see that the values of y no longer form an arithmetic progression.

Let us add another column to the table of values of argument and function (Fig. 2). In this column we shall write down by how much the value of y changes when the argument x passes from one of its values to the next. For example, suppose the argument changes from the value $x = 2$ to the value $x = 3$. Then the function changes from the value $y = 4$ to the value $y = 9$. The change, or, as they say, the *increment* of the function* equals the difference between the new value and the previous value of the function, namely,

$$9 - 4 = 5.$$

Thus in the third column of our table we write the increments of the function $y = x^2$. It can now be clearly seen that the function $y = x^2$ varies in such a way that if x increases, so does not only the function itself but also its increment. In the graph this fact is also apparent: the curve $y = x^2$ goes up more and more steeply, while the graph of the linear function, which varies uniformly, always forms the same angle with the x-axis (Fig. 3).

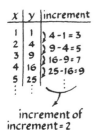

increment of
increment = 2
constant!

Fig. 2

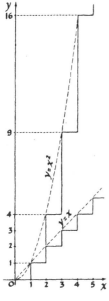

Fig. 3

*An increment of the function $y = f(x)$ is usually denoted by the Greek letter Δ (delta):

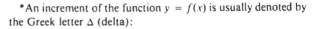

Δy, or $\Delta f(x)$.

40

It is interesting to note that the increments of the function $y = x^2$ form an arithmetic progression. The reader is asked to try to prove this fact in general: If the values of the argument x form an arithmetic progression,

$$a, a + d, a + 2d, \cdots, a + bd, \cdots,$$

then the values of the corresponding increments of the quadratic function $y = x^2$ also form an arithmetic progression.

If the argument t is the time and the function s is the distance covered (we change x to t and y to s according to the accepted notation in physics), then the relationship $s = t^2$ corresponds to uniformly accelerated motion (with acceleration equal to 2), while the formula $s = kt + b$ represents uniform motion with velocity k. In uniform motion a body covers equal distances in equal intervals of time; that is, equal increments in the function correspond to equal increments in the argument (a linear function converts every arithmetic progression into an arithmetic progression). In uniformly accelerated motion the distances of the path covered in equal intervals of time increase uniformly. Correspondingly, for a quadratic function (by the way, not only for $y = x^2$, but for any function $y = ax^2 + bx + c$), uniformly increasing increments of the functions correspond to equal increments of the argument.

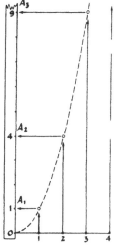

Fig. 4

EXERCISES

1. Work out a table with three columns (values of the argument, values of the function, and values of the increments of the function) for the trinomial $y = x^2 + x - 3$, taking for x the values $1, 0, -1, -2, -3$. Add one more column to the table and enter in it the differences between two consecutive increments.

Now take another trinomial: $x^2 + 3x + 5$. Work out a similar table for it. Compare the last columns

41

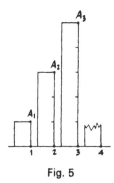

Fig. 5

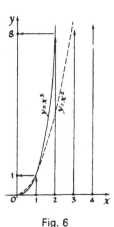

Fig. 6

of these two tables. And what is the result if the trinomial $y = 2x^2 + 3x + 5$ is used?

2. From Fig. 4 it is obvious that if a uniform scale is taken on the positive half of the x-axis, then the graph of $y = x^2$ transforms it into the scale O, A_1, A_2, A_3, etc., located on the y-axis, which is no longer uniform. With this scale the y-axis splits into the segments OA_1, A_1A_2, A_2A_3, etc. Imagine that the y-axis is cut into these segments, which are then placed vertically one after the other along the x-axis at equal distances from each other (with their base points at the points 1, 2, 3, etc.) (Fig. 5). How are the end points of the segments arranged? Explain the result.

3. Let us consider the graph of $y = x^3$ for positive values of x (Fig. 6). Do the same with it as with the graph of $y = x^2$ in Exercise 2. In your drawing, show how the end points of the segments are arranged in this case. A more difficult question: Can you find the equation of the curve on which the end points of the segments lie?

Problem

Construct the graph of $y = x^2$. Take a fairly large scale: $1 = 2$ cm (4 squares). Mark the point $F(0, \frac{1}{4})$ on the y-axis.

Measure the distance from the point F to any point M of the parabola with a strip of paper. Then fasten the strip at the point M and turn it around this point so that it becomes vertical. The end of the strip drops somewhat below the x-axis. Mark on the strip how much it goes beyond the x-axis (Fig. 7). Now take another point on the parabola and repeat the measurement. By how much does the edge of the strip now drop below the x-axis? We can tell you the result in advance: Whatever point is taken on the parabola $y = x^2$, the distance from this point to the point $(0, \frac{1}{4})$ will always be larger than the distance from the same point to the x-axis by the same number, $\frac{1}{4}$.

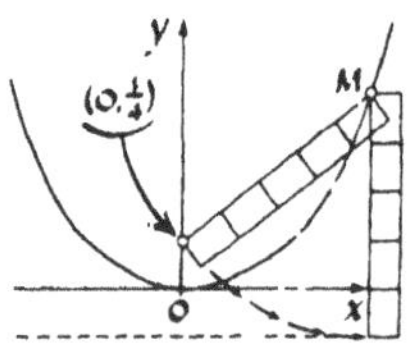

Fig. 7

42

This can be put differently: The distance from any point of the parabola to the point $(0, \frac{1}{4})$ is equal to the distance from the same point of the parabola to the line $y = -\frac{1}{4}$, which is parallel to the x-axis.

This remarkable point, $F(0, \frac{1}{4})$, is called the *focus* of the parabola $y = x^2$, and the straight line $y = -\frac{1}{4}$ is called the *directrix* of this parabola.

Every parabola has a directrix and a focus. (See also Fig. 1 on p. 44.)

3

Let us now consider the graphs of quadratic trinomials of the form

$$y = x^2 + px + q.$$

Let us show that in their form they differ in no way from the parabola $y = x^2$ but are only in a different position with respect to the coordinate axes.

To begin with, let us consider a numerical example. We take the trinomial

$$y = x^2 + 2x + 3.$$

To obtain its graph, let us represent it in the form

$$y = (x + 1)^2 + 2,$$

where we have completed the square.

The graph of $y = (x + 1)^2$ is obtained from the parabola $y = x^2$ by translation along the x-axis. (Explain why the curve $y = (x + 1)^2$ is obtained from $y = x^2$ by a translation to the left.) From the graph of $y = (x + 1)^2$ the graph of $y = (x + 1)^2 + 2$ can be obtained very simply (Fig. 8).

Thus, the graph of the trinomial

$$y = (x + 1)^2 + 2 = x^2 + 2x + 3$$

is obtained from the parabola $y = x^2$ by translation to the left by one unit and up by two units. In this

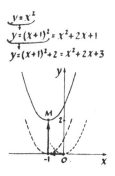

$y = x^2$

$y = (x+1)^2 = x^2 + 2x + 1$

$y = (x+1)^2 + 2 = x^2 + 2x + 3$

Fig. 8

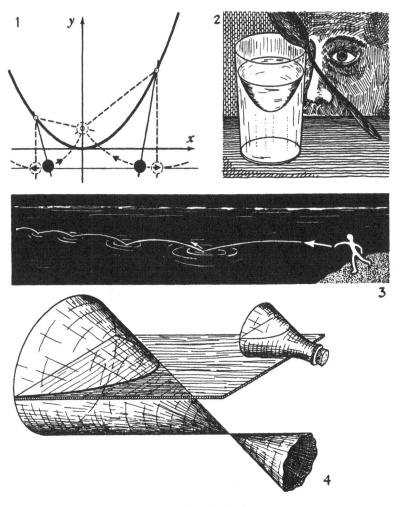

Interesting Properties of the Parabola

1. Any point of the parabola is equidistant from some point, called the focus of the parabola, and some straight line, called the directrix.

2. The surface of a liquid in a rotating container has the shape of a paraboloid of revolution. You can see this surface if you stir a partially filled glass of water vigorously with a small spoon and then remove the spoon.

3. If a stone is thrown at some angle to the horizon, then it travels along a parabola.

4. If the surface of a cone is intersected with a plane parallel to one of its generators, then a parabola is obtained as section.

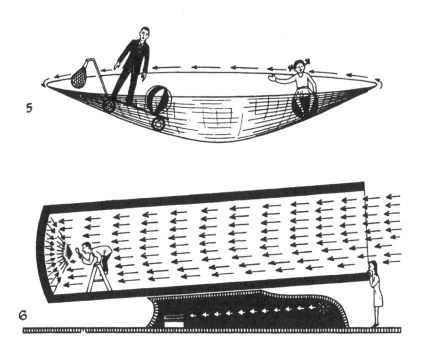

5. In parks an amusement attraction, "the miraculous paraboloid," is occasionally set up. To each of the persons standing in the paraboloid it seems that he is standing on the floor, while the other persons by some miracle stick to the walls.*

6. Parabolic mirrors are also used in reflecting telescopes: the light of the distant star, traveling in a parallel pencil, having fallen on the mirror of the telescope, converges to the focus.

*The experiments described in points 2 and 5 are based on the same property of the paraboloid: is rotated with appropriate speed about its vertically directed axis, then the resultant of the centrifugal force of gravitation in any point of the paraboloid is directed perpendicular to its surface.

transformation the vertex of the parabola, located at the point $(0, 0)$, the origin, goes into the point M with coordinates $(-1, 2)$.

EXERCISES

1. Draw graphs of the functions:

(*a*) $y = (x + 2)^2 + 3$;

(*b*) $y = (x + 2)^2 - 3$;

(*c*) $y = (x - 2)^2 + 3$;

(*d*) $y = (x - 2)^2 - 3$.

2. (*a*) Find the least value of the function

$$y = x^2 + 6x + 5.$$

Solution. The least value of the given function is the ordinate of the vertex of the parabola $y = x^2 + 6x + 5$. To determine the coordinates of the vertex, let us complete the square:

$$x^2 + 6x + 5 = (x + 3)^2 - 4.$$

Thus our parabola was obtained from $y = x^2$ by translation along the x-axis by -3 units and along the y-axis by -4 units; that is, the least value of the function equals -4.

(*b*) The vertex of the parabola $y = x^2 + px + q$ is located at the point $(-1, 2)$. Find p and q.

Now let us prove that by translating the parabola $y = x^2$, we can obtain the graph of any quadratic trinomial of the form

$$y = x^2 + px + q.$$

For this purpose, we complete the square, as before, that is, we represent our trinomial in the form $y = (x + \cdots)^2 + \cdots$, where the second summand in brackets and the constant term must be chosen so as not to depend on x.

After expansion of the expression in parentheses, the term linear in x is the result of doubling a product, and since this term must equal px, the second summand in the parentheses must be taken equal to $p/2$. Thus we have

$$x^2 + px + q = \left(x + \frac{p}{2}\right)^2 + \cdots$$

$$= x^2 + px + \frac{p^2}{4} + \cdots .$$

Since the constant term of the trinomial must equal q, instead of the three dots we have to take $q - p^2/4$.

Thus the trinomial $y = x^2 + px + q$ can be rewritten in the form

$$y = \left(x + \frac{p}{2}\right)^2 + q - \frac{p^2}{4} .$$

We see (Fig. 9) that *the graph of*

$$y = x^2 + px + q$$

represents the parabola $y = x^2$ translated* by $-p/2$ along the x-axis and by $q - p^2/4$ along the y-axis.

The vertex M of this parabola has the abscissa $x_M = -p/2$ and the ordinate $y_M = q - (p^2/4)$.

$$y = x^2 + px + q$$
$$= \left(x + \frac{p}{2}\right)^2 + q - \frac{p^2}{4}$$

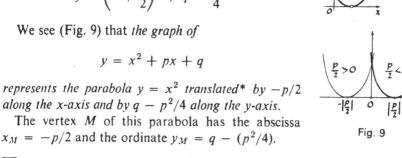

Fig. 9

4

Taking as "base" the graph of $y = ax^2$, in the same way we can obtain the graph of the quadratic trinomial of the more general form,

$$y = ax^2 + bx + c.$$

Let us analyze this, using an example. Let us take the trinomial $y = \frac{1}{2}x^2 - 3x + 6$. Let us put the coefficient of x^2 in front of the parentheses:

$$\tfrac{1}{2}x^2 - 3x + 6 = \tfrac{1}{2}(x^2 - 6x + 12).$$

*Translation by $-p/2$ along the x-axis means a translation to the right if $-p/2 > 0$, and a translation to the left if $-p/2 < 0$.

We complete the square in the expression inside the parentheses:

$$\tfrac{1}{2}(x^2 - 6x + 12) = \tfrac{1}{2}(x^2 - 2 \times 3x + 9 + 3)$$
$$= \tfrac{1}{2}[(x - 3)^2 + 3].$$

Thus, finally,

$$y = \tfrac{1}{2}(x - 3)^2 + \tfrac{3}{2}.$$

We see that the graph of $y = \tfrac{1}{2}x^2(x - 3)^2 + \tfrac{3}{2}$ is obtained from the parabola $y = \tfrac{1}{2}x^2$ by translating 3 units to the right along the x-axis and $\tfrac{3}{2}$ units up along the y-axis.

Problems

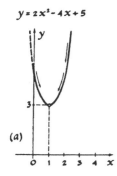

$y = 2x^2 - 4x + 5$

(a)

1. Translate the parabola ax^2 along the x-axis and y-axis so that the graph of the trinomial $y = ax^2 + bx + c$ is obtained.

(**Answer.** The parabola $y = ax^2 + bx + c$ is obtained from the parabola $y = ax^2$ by translation of $-b/2a$ along the x-axis and $(4ac - b^2)/4a$ along the y-axis.)

2. Find the least value of the function $y = 2x^2 - 4x + 5$ in the intervals:

(a) from $x = 0$ to $x = 5$ $\qquad (0 \leq x \leq 5)$,

(b) from $x = -5$ to $x = 0$ $\quad (-5 \leq x \leq 0)$.

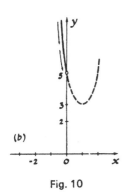

(b)

Fig. 10

Solution. Let us make use of the results of the preceding problem and construct the graph of the function $y = 2x^2 - 4x + 5$ (Fig. 10).

From the diagram it is obvious that as x varies from the value $x = 0$ to the value $x = 5$, the function $y = 2x^2 - 4x + 5$ initially decreases (to $x = 1$), and then increases. Thus the least value of the function $y = 2x^2 - 4x + 5$ in interval (a) is its value at $x = 1$ (Fig. 10a). As x varies from -5 to 0, the function $y = 2x^2 - 4x + 5$ constantly decreases. Hence

48

the least value of the function in interval (*b*) is its value at $x = 0$ (Fig. 10b).

(**Answer.** The least value in the interval (*a*) equals 3; in interval (*b*) it is 5.)

EXERCISES

1. Draw graphs of the following functions, indicating the exact coordinates of the vertex of each of the parabolas and the coordinates of the points of intersection of the graphs with the coordinate axes:

(*a*) $y = x - x^2 - 1$;

(*b*) $y = -3x^2 - 2x + 1$;

(*c*) $y = 10x^2 - 10x + 3$;

(*d*) $y = 0.125x^2 + x + 2$.

2. The graph of what function is obtained if the parabola $y = x^2$ is first stretched in the ratio 2:1 along the *y*-axis and then translated 3 units down along the same axis? The graph of what function is obtained if these two transformations are carried out in reverse order: first the parabola $y = x^2$ is translated 3 units downward, and then the curve thus obtained is expanded in the ratio 2:1 along the *y*-axis (Fig. 11)?

3. By how much must the parabola

$$y = x^2 - 3x + 2$$

be translated along the *x*-axis and along the *y*-axis so that the parabola $y = x^2 + x + 1$ is obtained?

4. Translate the parabola $y = x^2$ along the *x*-axis so that it will pass through the point (3, 2). The graph of what function is obtained (Fig. 12)?

 5

Let us now see what can be said about the solution of the quadratic equation $x^2 + px + q = 0$ using the graph of the function $y = x^2 + px + q$.

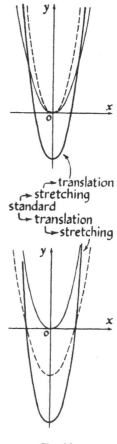

translation
stretching
standard
translation
stretching

Fig. 11

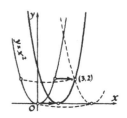

Fig. 12

49

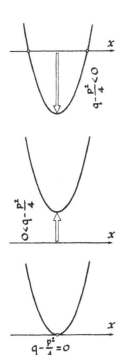

Fig. 13

The roots of this equation are those values of x for which the value of the function $y = x^2 + px + q$ is equal to zero. In the graph these points have ordinates equal to zero; that is, they lie on the x-axis.

From the graph of the quadratic trinomial

$$y = x^2 + px + q,$$

it is immediately obvious that the quadratic equation $x^2 + px + q = 0$ has two real roots if $p^2/4 - q > 0$, and has no roots if $p^2/4 - q < 0$. (Recall that the parabola $y = x^2$ moves down if $q - p^2/4 < 0$, and up if $q - p^2/4 > 0$) (see Fig. 13).

If $p^2/4 - q = 0$, our quadratic equation

$$x^2 + px + q = 0$$

transforms into the equation $(x + p/2)^2 = 0$. This case is particularly interesting. Let us consider it in detail.

The equation $x - 2 = 0$ has one solution, $x = 2$. The equation $(x - 2)^2 = 0$ also has but one solution, $x = 2$; no other number satisfies this equation.

However, in the first case we say that the equation $x - 2 = 0$ has one root, while in the second case we say that the equation $(x - 2)^2 = 0$ has a multiple root or two equal roots: $x_1 = 2$ and $x_2 = 2$.

How can this difference be explained?

There are several methods, and we shall give one of them. Let us alter the first equation a little: let us replace zero in the right-hand term by some small number. The root changes, of course, but it will remain unique, as before; the equation is satisfied by only one number, as before. For example,

$$x - 2 = 0.01, \qquad x = 2.01.$$

Let us now alter the second equation in the same way:

$$(x - 2)^2 = 0.01, \qquad x^2 - 4x + 3.99 = 0.$$

50

The resulting equation will now have two roots, $x_1 \approx 2.1$ and $x_2 \approx 1.9$. Now we shall again change the right-hand side in the equation $(x - 2)^2 = 0.01$, replacing it by smaller and smaller numbers. As long as this right-hand side does not equal zero, the equation will have two different roots. As the right-hand side diminishes, the roots "approach each other" so that their values will differ from one another by a smaller and smaller quantity. Finally, when the right-hand side becomes equal to zero, the two roots "coincide" — the values of the two roots become equal to each other. Therefore one says that the equation $(x - 2)^2 = 0$ has two roots, merging into a double root.

Geometrically the case of coincident roots corresponds to the parabola $y = (x - 2)^2$ touching the x-axis.

The general case of a quadratic trinomial $y = x^2 + px + q$ will be analyzed geometrically. Suppose, to begin with, the constant term q is smaller than $p^2/4$ (i.e., $q - p^2/4 < 0$), so that the parabola $y = x^2 + px + q$ has two points of intersection with the x-axis (Fig. 14). We shall increase the constant term: At first the parabola, moving up, will have two points of intersection with the x-axis (the equation $x^2 + px + q$ then has two distinct roots); then as these points of intersection get closer, at a certain moment (when $q - p^2/4 = 0$) they merge into one point. At this moment the parabola

$$y = x^2 + px + q = \left(x + \frac{p}{2}\right)^2$$

touches the x-axis, and the equation

$$x^2 + px + \frac{p^2}{4} = 0$$

has one double root. As the constant then is increased further, the parabola ceases to intersect the x-axis, and the equation $x^2 + px + q = 0$ will not have any real roots.

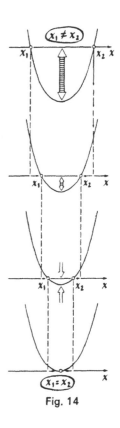

Fig. 14

EXERCISES

1. Find the parabola $y = ax^2 + bx + c$ which intersects the x-axis at the points $x = 3$ and $x = -5$, and the y-axis at the point $y = 30$.

Solution. The quadratic trinomial defining this parabola will have the form

$$a(x - 3)(x + 5).$$

The point of intersection with the y-axis is obtained by setting $x = 0$. Hence, when $x = 0$ our function must equal 30. We obtain

$$a(-3)(+5) = 30, \qquad \text{hence } a = -2.$$

(**Answer.** The parabola $y = -2x^2 - 4x + 30$.)

2. (*a*) Find the quadratic trinomial of the form $x^2 + px + q$, if its graph intersects the x-axis at the points $x = 2$ and $x = 5$.

(*b*) Find the cubic polynomial of the form $y = x^3 + px^2 + qx + r$, if it is known that its graph intersects the x-axis at the points $x = 1$, $x = 2$, and $x = 3$.

(*c*) Can you devise a polynomial whose graph would intersect the x-axis at 101 points: $x_1 = -50$, $x_2 = -49$, $x_3 = -48, \cdots, x_{101} = 50$?

What is the least degree of such polynomials?

3. The trinomial $-x^2 + 6x - 9$ has two identical roots:

(*a*) Change the constant term by 0.01 so that the resulting trinomial has two distinct roots.

(*b*) Can the same result be obtained by making a change of 0.01, only this time in the coefficient of x?

4. Figures 15a and b represent graphs of quadratic trinomials $y = x^2 + px + q$. Find p and q. Draw the graph of Fig. 15b, making a more fortunate choice of scale and position of the axes.

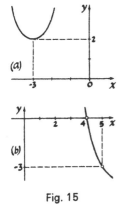

Fig. 15

52

5. Figures 16a, b, and c depict graphs of the quadratic trinomials $y = ax^2 + bx + c$. Find a, b, and c.

6. (a) Solve the inequality

$$x^2 - 5x + 4 > 0.$$

Solution. From Fig. 17 it is obvious that the function $y = x^2 - 5x + 4$ is positive in two intervals: for x less than 1 and for x larger than 4.

(**Answer.** $x < 1$ and $x > 4$.)

(b) Solve the inequality

$$x - 1 < |x^2 - 5x + 4|.$$

Solution. Let us draw in one diagram the functions on the right-hand and left-hand sides. From Fig. 18 it is apparent that the straight line $y = x - 1$ has three points in common with the graph of

$$y = |x^2 - 5x + 4|: A(x_1, y_1), B(x_2, y_2), C(x_3, y_3).$$

The condition $x - 1 < |x^2 - 5x + 4|$ is satisfied in three intervals: $x < x_1$, $x_1 < x < x_2$, $x > x_3$. The values of x_1 and x_3 can be found from the equation

$$x - 1 = x^2 - 5x + 4.$$

The value of x_2 is found from the equation

$$x - 1 = -(x^2 - 5x + 4).$$

(**Answer.** $x < 1$, $1 < x < 3$ and $x > 5$, that is, all x, except $x = 1$ and $3 \le x \le 5$.)

(c) Write down the answer for the inequalities

$$x - 1 > |x^2 - 5x + 4|,$$
$$x - 1 \ge |x^2 - 5x + 4|.$$

53

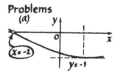

Problems
(a)

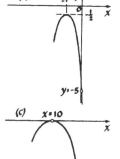

(b)

(c) $x = 10$

(12,-4)

Fig. 16

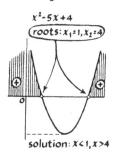

$x^2 - 5x + 4$
(roots: $x_1 = 1, x_2 = 4$)

solution: $x < 1, x > 4$

Fig. 17

Problem
$x - 1 < |x^2 - 5x + 4|$
($y = |x^2 - 5x + 4|$)

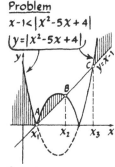

Solution: $x < x_1$,
$x_1 < x < x_2$,
$x_3 < x$

Fig. 18

7. Find the largest value of the function

$$y = x^2 - 5|x| + 4$$

in the interval from -2 to 2.

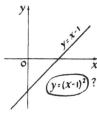

Fig. 19

6

The graph of $y = x^2$ can also be drawn by "squaring" the graph of $y = x$, that is, mentally squaring the value of each ordinate (Fig. 19).

EXERCISES

1. Given the graph (Fig. 20) of $y = x - 1$, draw the graph of $y = (x - 1)^2$ in the same diagram.

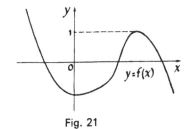

Fig. 20   Fig. 21

2. Given the graph (Fig. 21) of $y = f(x)$, draw the graph of $y = (f(x))^2$ in the same diagram.

3. Making use of the graph of

$$y = x(x + 1)(x - 1)(x - 2)$$

(Fig. 9 on p. 14), draw the graph of

$$y = x^2(x + 1)^2(x - 1)^2(x - 2)^2.$$

4. Draw graphs of:

(a) $y = [x]^2$,

(b) $y = (x - [x])^2$.

54

CHAPTER 5

The Linear Fractional Function

1

Figure 1 represents the "graph" of the function $y = 1/x$ in the form in which it is frequently drawn by persons not initiated into the construction of graphs. They argue like this: "For $x = 1$, $y = 1$. For $x = 2$, $y = \frac{1}{2}$. For $x = 3$, $y = \frac{1}{3}$. For $x = -1$, $y = -1$. For $x = 0, \cdots$? It's unclear.... It is not known what $1/0$ means, and therefore we omit $x = 0 \cdots$."

The reader knows from the preceding text that graphs must not be drawn in this manner. In order to get a correct picture, let us note first that at $x = 0$ the function is not defined. In such cases it is interesting to see how the function behaves near this point. When x, decreasing in absolute value, approaches zero, then y becomes as large in absolute value as we please. If x approaches zero from the right ($x > 0$), then $y = 1/x$ is also positive. Therefore, approaching zero from the right, the curve of the graph moves up (Fig. 2a). If x approaches zero from the left ($x < 0$),

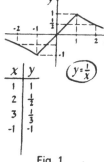

x	y
1	1
2	$\frac{1}{2}$
3	$\frac{1}{3}$
-1	-1

$\left(y = \frac{1}{x}\right)$

Fig. 1

$x > 0$, $y = \frac{1}{x} > 0$

Let $x \to 0$

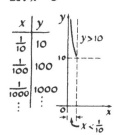

x	y
$\frac{1}{10}$	10
$\frac{1}{100}$	100
$\frac{1}{1000}$	1000
$\vdots$	$\vdots$

$y > 10$

$x < \frac{1}{10}$

Fig. 2a

$$y = f(x) \rightarrow y = \frac{1}{f(x)}$$

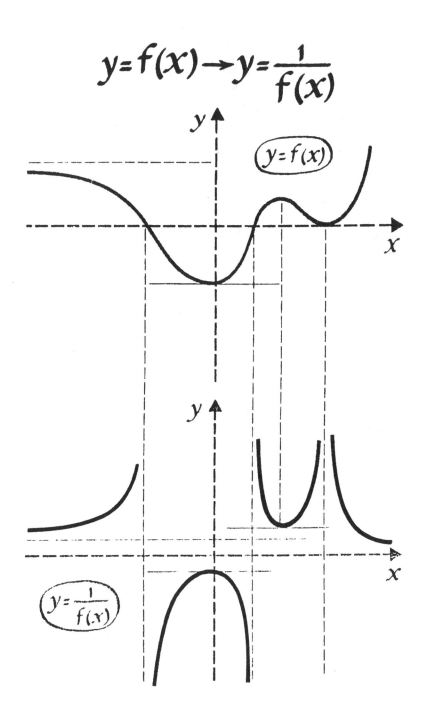

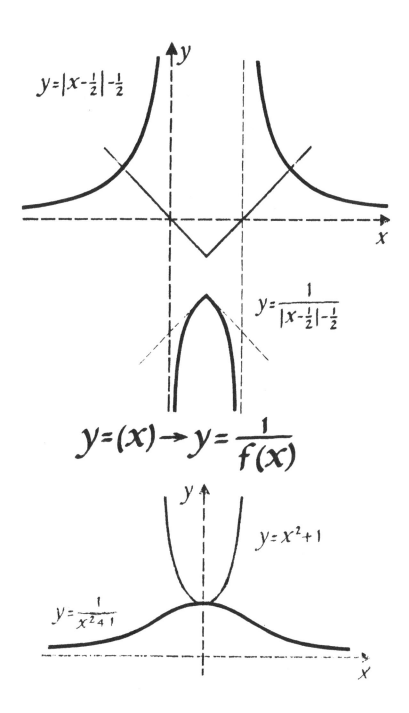

$$y = |x - \tfrac{1}{2}| - \tfrac{1}{2}$$

$$y = \frac{1}{|x - \tfrac{1}{2}| - \tfrac{1}{2}}$$

$$y = (x) \rightarrow y = \frac{1}{f(x)}$$

$$y = x^2 + 1$$

$$y = \frac{1}{x^2 + 1}$$

Let $x \to 0$ for $x < 0$

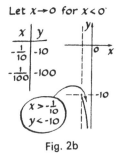

x	y
$-\frac{1}{10}$	-10
$-\frac{1}{100}$	-100

$x > -\frac{1}{10}$
$y < -10$

Fig. 2b

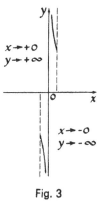

$x \to +0$
$y \to +\infty$

$x \to -0$
$y \to -\infty$

Fig. 3

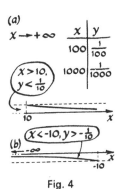

(a)
$x \to +\infty$

x	y
100	$\frac{1}{100}$
1000	$\frac{1}{1000}$

$x > 10$,
$y < \frac{1}{10}$

(b) $x < -10, y > -\frac{1}{10}$

$\to -\infty$

Fig. 4

then y is negative, and therefore from the left the graph moves down (Fig. 2b).

It is now obvious that near the "forbidden" value, $x = 0$ the graph, having separated into two branches, diverges along the y-axis: the right branch goes up, while the left goes down (Fig. 3).

Let us find out now how the function behaves if x increases in absolute value. First let us consider the right branch, that is, values of $x > 0$. For positive x the values of the function y are also positive. This means that the entire right branch is above the x-axis. As x increases, the fraction $1/x$ decreases. Therefore, in moving from zero to the right, the curve $y = 1/x$ drops lower and lower, and it can approach the x-axis to within an arbitrarily small distance (Fig. 4a). For $x < 0$ an analogous picture is obtained (Fig. 4b).

Thus as x increases indefinitely in absolute value, the function $y = 1/x$ decreases indefinitely in absolute value and both branches of the graph approach the x-axis: the right-hand one from above, the left from below (Fig. 5).

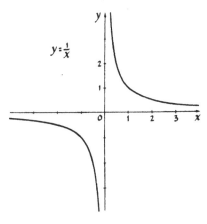

$y = \frac{1}{x}$

Fig. 5

The curve that is the graph of $y = 1/x$ is called a *hyperbola*. The straight lines that the branches of the hyperbola approach are called its *asymptotes*.

58

The graph of $y = 1/x$ can be constructed somewhat differently.

Let us draw the graph of the function $y = x$ (Fig. 6a). Let us replace each ordinate by its inverse and draw the corresponding points in Fig. 6b. We thus obtain the graph of $y = 1/x$.

The picture we have drawn clearly shows how small ordinates of the first graph transform into large ordinates of the second, and, on the other hand, large ordinates of the first transform into small ordinates of the second.

This method of "dividing" graphs is useful whenever we can construct the graph of $y = f(x)$, and we have to construct the graph of the function $y = 1/f(x)$ (see pp. 56 and 57).

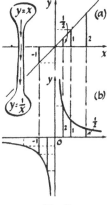

Fig. 6

EXERCISES

1. From the graph of $y = x^2$ construct the graph of $y = 1/x^2$. (Solution in Fig. 7.)

2. Construct the graphs of

(a) $y = \dfrac{1}{x^2 - 3x - 2}$;   (b) $y = \dfrac{1}{x^2 - 2x + 3}$.

(It will be seen that these two graphs look quite different.)

3. From the graph of $y = [x]$ (see p. 6) and $y = x - [x]$, construct graphs of

(a) $y = \dfrac{1}{[x]}$;   (b) $y = \dfrac{1}{x - [x]}$.

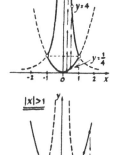

Fig. 7

The curves that you are asked to construct in the following exercises are obtained from the hyperbola $y = 1/x$ by transformations with which you are already familiar. All of them are also called hyperbolas.

1. Draw the graphs of the functions:

(*a*) $y = 1/x + 1$;

(*b*) $y = 1/(x + 1)$;

(*c*) $y = 1/(x - 2) + 1$.

Indicate which are the asymptotes of each of these hyperbolas.

2. (*a*) Prove that the straight lines $y = x$ and $y = -x$ are axes of symmetry of the hyperbola $y = 1/x$.

(*b*) Does the right-hand branch of the graph of $y = 1/x^2$ have an axis of symmetry? $\oplus$

3. From the graph of $y = 1/x$ construct the graph of $y = 4/x$. Does this curve have any axes of symmetry?

The graphs of functions of the form

$$y = \frac{b}{cx + d} \qquad \text{(where } c \neq 0 \text{ and } b \neq 0\text{)}$$

can be obtained from the graph of $y = 1/x$ by translation along the x-axis and stretching along the y-axis. In order to determine the correct value of the translation and the stretching ratio, it is necessary to divide the numerator and denominator of the fraction by c, the coefficient of x:

$$\frac{b}{cx + d} = \frac{b/c}{x + (d/c)}.$$

Let us do this for the example $y = 1/(3x + 2)$. We have (Fig. 8)

$$\frac{1}{3x + 2} = \frac{\frac{1}{3}}{x + \frac{2}{3}}.$$

It is now obvious that the graph of our function $y = 1/(3x + 2)$ is the graph of $(1/x)$, translated by $(-\frac{2}{3})$* along the x-axis and contracted along the y-axis in the ratio $3:1$ (Fig. 8).

*Not by (-2), as some students would say rashly.

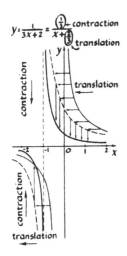

$y = \dfrac{1}{3x+2} = \dfrac{\frac{1}{3}}{x+\frac{2}{3}}$ contraction

translation

contraction

translation

contraction

translation

Fig. 8

Draw the graph of

$$y = \frac{1}{2-x} + 1.$$

(**Hint.** Transform the fraction $1/(2-x)$ as already stated: Divide numerator and denominator by the coefficient of x, that is, by (-1), to obtain

$$y = \frac{-1}{x-2} + 1.)$$

4

The graphs of functions of the form

$$y = \frac{ax+b}{cx+d},$$

called *linear fractional functions*, do not differ in form from the graph of $y = 1/x$. We assume, of course, that $c \neq 0$ (otherwise the linear function $y = (a/d)x + b/d$ is obtained) and that $a/c \neq b/d$; that is, the numerator is not a multiple of the denominator (as with the function

$$y = \frac{4x+6}{2x+3}),$$

otherwise the function is constant.

Let us prove this. We first consider the example: $y = (2x+1)/(x-3)$. Let us separate the "integral part" of the fraction, dividing the numerator by the denominator (Fig. 9). We obtain

$$\frac{2x+1}{x-3} = 2 + \frac{7}{x-3}.$$

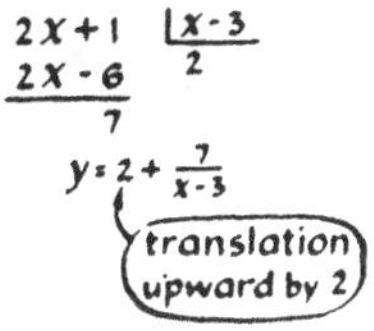

$$\begin{array}{r} 2x+1 \\ 2x-6 \\ \hline 7 \end{array} \begin{array}{|l} \underline{x-3} \\ 2 \end{array}$$

$$y = 2 + \frac{7}{x-3}$$

translation
upward by 2

Fig. 9

The graph of this function is evidently obtained from the graph of $y = 1/x$ by the following transformations: a translation by 3 units to the right, a stretching in the ratio $7:1$ along the y-axis, and a translation by 2 units upward.

61

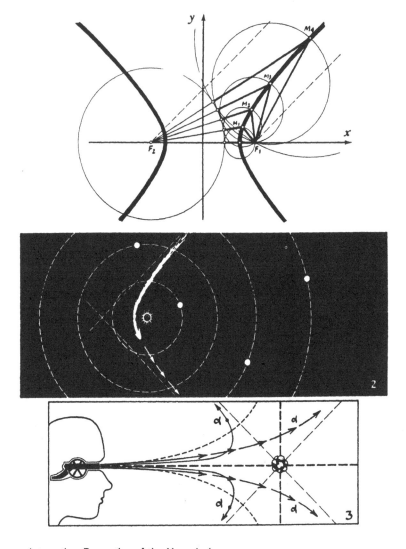

Interesting Properties of the Hyperbola

1. The hyperbola is the locus of all points M, the difference of whose distances from two given points F_1 and F_2, called foci, is equal in absolute value to a given number.

2. A comet or meteorite traveling into the solar system from a great distance moves on a branch of a hyperbola, with the sun in its focus. One asymptote* gives the direction in which the comet approaches, and the second asymptote gives the direction in which it leaves the solar system.

3. In the bombardment of an atomic nucleus, an α-particle, flying past the nucleus, travels in a hyperbola.

*Every hyperbola has two asymptotes. The hyperbolas that are graphs of the linear fractional function $y = (ax + b)/(cx + d)$ have mutually perpendicular asymptotes. Other hyperbolas have asymptotes intersecting each other at a different angle.

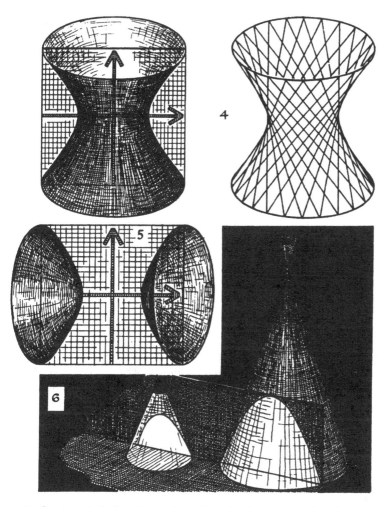

4. If a hyperbola is rotated about its axis of symmetry that does not intersect its branches, a surface is obtained called a *hyperboloid of one sheet*. This surface has a striking property: it is "woven" from straight lines. The tower of the Moscow telecenter is composed of "pieces" of such hyperboloids, made entirely from straight steel rods.

5. If a hyperbola is rotated about the other axis of symmetry, a surface consisting of two "pieces," the *hyperboloid of two sheets* is obtained. It was this hyperboloid that A. Tolstoi had in mind in his novel, *The Hyperboloid of Engineer Garin*. But then, the property required by engineer Garin, to collect light rays in a parallel pencil, is actually not possessed by the hyperboloid but the paraboloid, so that it would be more correct to call the book *The Paraboloid of Engineer Garin*.

6. If an infinite cone is appropriately intersected with a plane, then at the intersection a hyperbola is obtained. If the reader has a lamp with a shade in the form of a circular cone, he can satisfy himself of the truth of this: the lamp illuminates a part of the wall, which is bounded by a piece of a hyperbola.

Any fraction $y = (ax + b)/(cx + d)$ can be written in an analogous way, separating its "integral part." Consequently, the graphs of all linear fractional functions $y = (ax + b)/(cx + d)$ are hyperbolas (translated different distances along the coordinate axes and stretched in different ratios along the y-axis).

Remark. To construct the graph of some linear fractional function, the fraction defining this function does not have to be transformed. Since we know that the graph is a hyperbola, it suffices to find the straight lines which its branches approach (the asymptotes of the hyperbola) and a few more points.

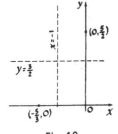

Fig. 10

Example. Let us construct the graph of the function

$$y = \frac{3x + 5}{2x + 2}.$$

Let us first find the asymptotes of this hyperbola. The function is not defined where $2x + 2 = 0$, that is, at $x = -1$ (Fig. 10). Consequently, the straight line $x = -1$ serves as vertical asymptote.

In order to find the horizontal asymptote, let us find what number the values of the function approach as the argument increases in absolute value. For large (in absolute value) values of x,

$$y = \frac{3x + 5}{2x + 2} \approx \frac{3x}{2x} = \frac{3}{2}.$$

Consequently, the horizontal asymptote is the straight line $y = \frac{3}{2}$.

Let us determine the points of intersection of our hyperbola with the coordinate axes. At $x = 0$ we have $y = \frac{5}{2}$. The function equals zero when $3x + 5 = 0$, that is, when $x = -\frac{5}{3}$.

Having marked the points $(-\frac{5}{3}, 0)$ and $(0, \frac{5}{2})$ in the diagram, we construct the graph (Fig. 11).

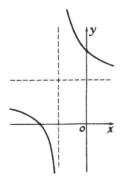

Fig. 11

64

EXERCISES

1. Construct graphs of the functions:

(a) $y = \dfrac{1}{1 - 2x}$; (b) $y = \dfrac{3 + x}{3 - x}$;

(c) $y = \left|\dfrac{2x + 1}{x + 1}\right|$.

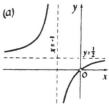

2. Figures 12a and b represent the graphs of linear fractional functions $y = (px + q)/(x + r)$. Find these functions (determine p, q, and r).

3. (a) How many solutions has the equation

$$\frac{x}{1 - x} = x^2 + 4x + 2?$$

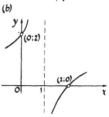

Fig. 12

Solution. Let us construct in one diagram the graphs of the functions

$$y = \frac{x}{1 - x} \quad \text{and} \quad y = x^2 + 4x + 2.$$

In Fig. 13 two points of intersection of these graphs can be seen. Evidently, there is also a third point, since the parabola intersects the asymptote of the hyperbola. The abscissas of the points of intersection of the graphs are the solutions of the equation.

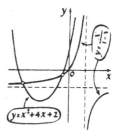

(Answer. Three solutions.)

Fig. 13

CHAPTER 6

Power Functions

1

Power functions are functions of the form $y = x^n$. We have already constructed graphs of the power functions for $n = 1$ and $n = 2$. For $n = 1$ we obtain the function $y = x$; the graph of this function is a straight line (Fig. 1a). For $n = 2$ the function $y = x^2$ is obtained; the graph of this function is a parabola (Fig. 1b).

The graph of the function $y = x^3$ ($n = 3$) is also called a parabola — a *parabola of the third degree* or *cubic parabola*. For positive values of the argument the cubic parabola $y = x^3$ is similar to the parabola of the second degree $y = x^2$. In fact, at $x = 0$ the function $y = x^2$ equals zero and the function $y = x^3$ also equals zero; both graphs pass through the origin; at $x = 1$ the value of x^2 equals 1 and that of x^3 also equals 1; and both graphs pass through the point $(1, 1)$.

As x increases (if x is positive), the values of the function $y = x^2$ as well as the values of the function $y = x^3$ increase. To the right of the origin the cubic

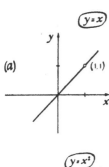

(a)

$\boxed{y = x}$

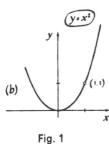

(b)

$\boxed{y = x^2}$

Fig. 1

66

parabola $y = x^3$, like the ordinary parabola $y = x^2$, steadily rises (Fig. 2).

For negative values of x the behavior of the curve $y = x^3$ is different from that of the curve $y = x^2$: for negative x the values of x^3 are also negative, therefore the curve heads downward (Fig. 3). Thus, on the whole, the cubic parabola is altogether different from the quadratic one.

The left half of the graph of $y = x^3$ can be obtained from its right half by using symmetry, of a different kind though from that which we considered on pages 10 and 11. Let us take any point M on the right half of the graph of $y = x^3$ (Fig. 3). If a is the abscissa of this point, then its ordinate b equals a^3 ($b = f(a) = a^3$). Let us now find the point of the graph that corresponds to the value of the abscissa with opposite sign, $x = -a$. The ordinate of such a point equals $(-a)^3$, that is, $-a^3$, or $-b$. Thus for each point $M(a, b)$ on the right half of the graph of $y = x^3$ there is a point $M'(-a, -b)$ on its left half. Obviously (Fig. 3), the point M' is symmetric to the point M with respect to the origin. Hence the entire left half of the graph can be obtained from the right half by symmetric reflection with respect to the origin of coordinates.

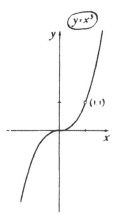

Fig. 2

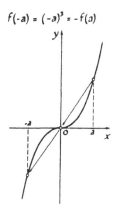

$f(-a) = (-a)^3 = -f(a)$

Fig. 3

EXERCISES

1. Which of the graphs of the following functions have a center of symmetry, which have an axis of symmetry:

$$y = x^4; \quad y = x^5; \quad y = x^7; \quad y = x^{16}?$$

2. Prove that the graph of the function $y = 1/x^3$ is symmetric with respect to the origin.

Solution. Let us consider the two points of the curve $y = 1/x^3$ with abscissas $x = a$ and $x = -a$. The ordinate of the first equals $1/a^3$, the ordinate of the second $1/(-a)^3 = -1/a^3$.

67

Consequently, for each point $M(a, 1/a^3)$ of our curve there is a point $M(-a, -1/a^3)$ symmetric to the first with respect to the origin. Therefore, the whole curve $y = 1/x^3$ is symmetric with respect to the origin.

3. Which of the following functions are even and which are odd:*

$$y = x^3|x|; \qquad y = |x^3| + x; \qquad y = \frac{x}{|x|};$$

$$y = |x - x^2|; \qquad y = (2x + 1)^4 + (2x - 1)^4;$$

$$y = \frac{1}{|2x - x^2|} - \frac{1}{|2x + x^2|};$$

$$y = (x^3 + 1)^2; \qquad y = (x^2 + 1)^3;$$

$$y = \cfrac{1}{x + \cfrac{1}{x + (1/x)}};$$

$$y = (3 - x)^5 - (3 + x)^5?$$

2

Let us now see how the graphs of $y = x^3$ and $y = x^2$ differ from each other for positive values of x. To this end let us represent x^3 as $x^2 \cdot x$ and obtain the graph of $y = x^3$, by "multiplying" the graph of $y = x^2$ by the graph of $y = x$ (Fig. 4). At $x = 1$ the value of x^3 equals the value of x^2: the point $(1, 1)$ is common to both graphs. Let us now go to the right and to the left from this point.

To the right of the point $(1, 1)$ the values of the function $y = x^3$ are obtained from the values of the function $y = x^2$ by multiplication by numbers larger than one. Therefore for $x > 1$ the values of x^3 are larger than x^2 — to the right of the point $(1, 1)$ the graph of the cubic parabola $y = x^3$ lies above the graph of the parabola $y = x^2$, and as the value of x

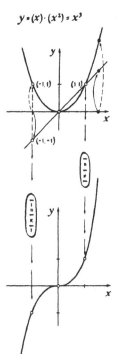

$y \cdot (x) \cdot (x^2) = x^3$

Fig. 4

*The definition of an even function is given on p. 11, the definition of an odd function is on p. 72.

68

increases, so does the difference between the two functions.

If one goes from the point (1, 1) to the left, toward $x = 0$, the values of x^3 will be obtained from the values of x^2 by multiplication by a number less than one. Therefore to the left of the point (1, 1) the cubic parabola lies below the parabola $y = x^2$, approaching the x-axis at the origin more quickly (Fig. 5).

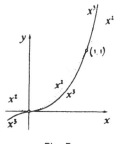

Fig. 5

QUESTIONS

For what x will the value of x^3 be 100 times larger than the value of x^2? 1000 times? How much above the quadratic parabola will the cubic parabola be at these x? Will the cubic parabola rise at any x 100 units above the quadratic parabola? 1,000,000?

If the thickness of a pencil line is considered to be equal to 0.1 mm and the unit of measure is taken to be 1 cm, at $x = 0.1$ the parabola $y = x^2$ can no longer be distinguished from the x-axis. How many times closer to the x-axis is the cubic parabola at this value of x?

 3

Obtaining x^3 from x^2 by multiplication by x, *how many times* is the ordinate of $y = x^3$ larger (or smaller) than the ordinate of $y = x^2$? Let us now try to represent graphically the difference between the values of the functions $y = x^3$ and $y = x^2$. To this end let us draw the graph of the function

$$y = x^3 - x^2,$$

whose ordinates can be obtained by subtracting from the ordinates of the graph of $y = x^3$ the ordinates of the graph of $y = x^2$ (Fig. 6).

At $x = 0$ both x^3 and x^2 vanish; consequently the graph of the function $y = x^3 - x^2$ passes through the origin. To the left of the origin the positive x^2 is subtracted from the negative x^3; the difference between $x^3 - x^2$ is negative so that the

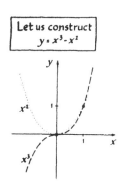

Fig. 6

69

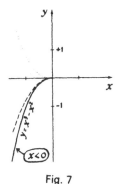

Fig. 7

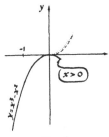

Fig. 8

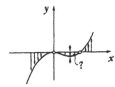

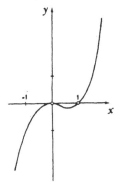

Fig. 9

Fig. 10

graph of $y = x^3 - x^2$ runs below the x-axis (and even below the graph of $y = x^3$ (see Fig. 7).

To the right of the origin things are more complicated; both functions are positive, and the result depends on which is larger in absolute value: x^3 or x^2. At first x^2 is larger than x^3; therefore the curve $y = x^3 - x^2$ is situated below the x-axis near the origin (Fig. 8). Gradually x^3 starts to grow faster and faster and at $x = 1$ catches up with the function $y = x^2$ in value. Therefore (somewhere between $x = 0$ and $x = 1$) the curve $y = x^3 - x^2$ starts to rise and at $x = 1$ intersects the x-axis (Fig. 9).

Further, after $x = 1$ the values of the function $y = x^3 - x^2$ increase, the graph moves up and for large x, when x^2 is small in comparison with x^3, is almost indistinguishable in its shape from the graph of $y = x^3$ (Fig. 10).

EXERCISES

1. It is possible to determine approximately at which x the values of the function $y = x^3 - x^2$ start to increase. Try to find this value of x with an accuracy of at least 0.1 (that is, one digit after the decimal point). We shall later be able to find the exact value. Exactly at this spot is also located the lowest point of the dip of the graph.

2. Solve the inequalities:

$$x^3 - x^2 > 0; \qquad x^3 - x^2 \leq 0.$$

Let us now compare the behavior of the functions $y = x^3$ and $y = cx^2$ and construct the graph of the function $y = x^3 - cx^2$. Let us first take a small value of c, for example, $c = 0.3$. The appearance of the graph depends on how the graph of $y = x^3$ and the graph of $y = 0.3x^2$ are situated with respect to each other. In Fig. 11 it is easy to understand that the construction of the graph far away from the origin, that is, for values of x of large absolute value, does not cause any difficulties. However, from the diagram it cannot be determined which of the parabolas $y = x^3$ and $y = 0.3x^2$ lies below the other near the origin, yet the answer to this question determines whether there will be a dip in the resulting graph or not.

In order to clarify this question, let us solve the inequality:

$$x^3 > 0.3x^2, \quad \text{or} \quad x^2(x - 0.3) > 0.$$

It is now clear that close to the origin, namely for positive values of x less than 0.3, the cubic parabola lies below the parabola $y = 0.3x^2$ (Fig. 12a). Therefore we can now draw the part of Fig. 11 which was not clear to us before in magnified form and construct the graph of the difference

$$y = x^3 - 0.3x^2.$$

In this graph, as in the graph of

$$y = x^3 - x^2,$$

there will be a dip, but a narrower one (Fig. 12b).

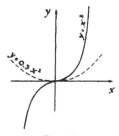

Fig. 11

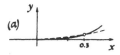

(a)

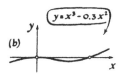

(b)

Fig. 12

EXERCISES

1. Find the width of the "dip" for the graphs of the functions:

(a) $y = x^3 - 0.01x^2$; (b) $y = x^3 - 1000x^2$.

2. Will there be a dip in the graph of

$$y = x^3 + 0.001x^2?$$

3. Show after which x the parabola $y = x^3$ will lie above the parabola $y = 50x^2$; the parabola $y = 10,000x^2$.

Having done these exercises, you will understand that the graphs of the functions $y = x^3 - cx^2$ for any $c > 0$ have the same character, the same form: to the left of the origin the graph goes downward,

$f(-a) =$	The function $y = f(x)$ is called odd if for each a the equation $f(-a) = -f(a)$ is satisfied.
$= -f(a)$	The graph of an odd function is symmetric with respect to the orgin.

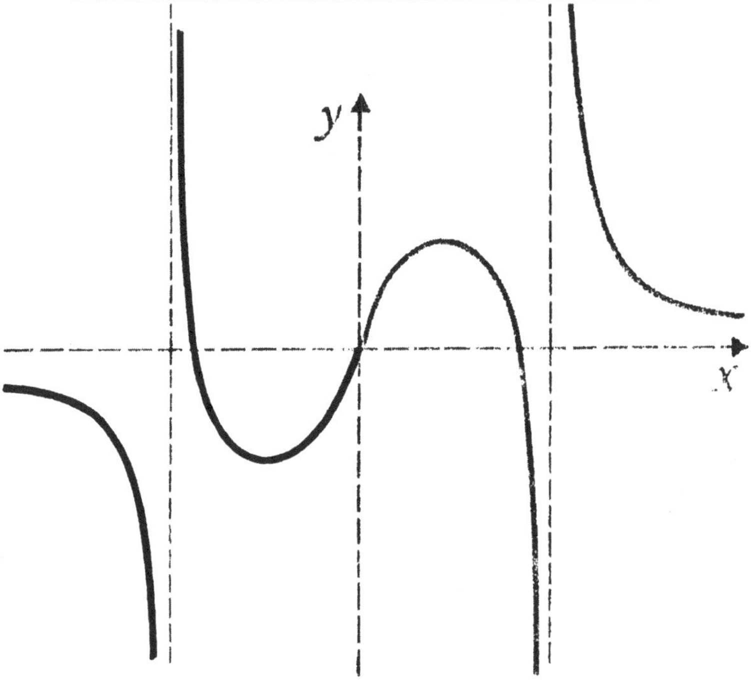

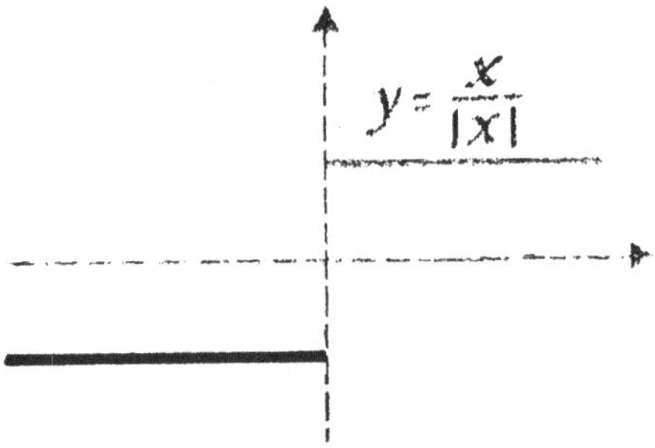

$y = x^3$

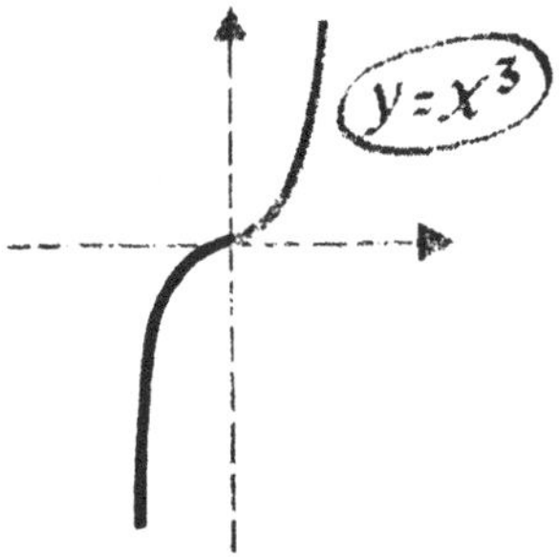

$y = \dfrac{x}{|x|}$

at the origin it touches the x-axis, farther on it turns downward again and then up: the dip in the graph that is obtained as a result becomes more pronounced as c becomes larger (Fig. 13a).

If we gradually diminish c, the dip will gradually flatten out and finally, when c becomes equal to zero, will disappear and the graph will turn into the ordinary cubic parabola $y = x^3$ (Fig. 13b).

4

We can now draw a general conclusion as to how the function $y = x^3$ behaves for positive values of x in comparison with any function of the form $y = cx^2$. For x close to zero, the function $y = x^3$ will be less than any function $y = cx^2$, even if the coefficient c is very small. For larger values of x, on the other hand, the function $y = x^3$ will be larger than any function of the form $y = cx^2$, even if the coefficient c is very large.

This can be expressed differently: the parabola of the third degree approaches the x-axis at the origin so quickly that between the parabola and the x-axis there can pass not only no straight line but also not one parabola $y = cx^2$, however small the coefficient c. On the other hand, for large values of x (for $x > 0$) the parabola $y = x^3$ "outruns" any parabola $y = cx^2$ for any coefficient c, no matter how large.

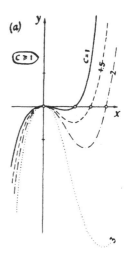

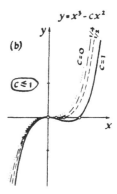

$y = x^3 - cx^2$

Fig. 13

EXERCISES

1. Draw graphs of the functions:

$$y = -x^3, \quad y = |x^3|; \quad y = 1 + x^3;$$
$$y = (2 + x)^3;$$
$$y = (2 - x)^3; \quad y = x^3 + 3x^2 + 3x.$$

2. Figure 14 represents the parabolas $y = 5x^3$ and $y = x^2$; because of the small scale of the diagram the relative position of the graphs near zero is not clear.

Look at this drawing "under a microscope" and draw what you see there: draw on a larger scale the region marked off by a small circle.

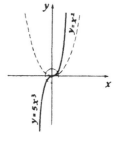

Fig. 14

73

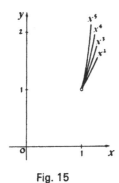

Fig. 15

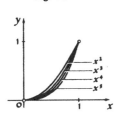

Fig. 16

5

We shall not analyze the functions $y = x^n$ for $n > 3$ as fully as $y = x^3$. The graphs of these functions remind one in their outward appearance either of the parabola $y = x^2$ (for n even), or of the parabola $y = x^3$ (for n odd).

It is understood that the function $y = x^4$ for large values (large in absolute value) of x grows still faster than $y = x^3$. Generally, the larger n is, the faster will grow the power function $y = x^n$ for large values of x (Fig. 15).

When x approaches zero, the values of all power functions $y = x^n$ also approach zero; the larger n is, the faster they will do so. The graphs of all power functions $y = x^n$ (starting with $n = 2$) touch the x-axis at the origin of coordinates, approaching it more rapidly the larger n is (Fig. 16).

For large n it is practically impossible to draw the graph of the function $y = x^n$ according to scale. On almost the whole segment from 0 to 1 the values of the function are very small and the graph of $y = x^n$ is indistinguishable from the x-axis. In a small region around $x = 1$, the function grows to 1 and then quickly grows at such a rate that the graph goes beyond the edges of any sheet of paper.* For example, let $n = 100$. Let us try to draw the graph of $y = x^{100}$, beginning with $x = 1$. At $x = 2$ we get $y = 2^{100}$. This is too big! Let us take $x = 1.1$. Then $y = (1.1)^{100}$. This is still a large number. In fact, $(1.1)^{100} = [(1.1)^{10}]^{10}$.

Let us now use the inequality $(1 + \alpha)^n > 1 + n\alpha$ (valid for $\alpha > 0$).† We obtain $(1.1)^{10} > 1 + 10 \times 0.1 = 2$. Thus, $(1.1)^{100} > 2^{10} > 1000$.

---

*For negative values of x the situation is analogous.

$$†(1 + \alpha)^n = (1 + \alpha)(1 + \alpha) \cdots (1 + \alpha)$$
$$= 1 + \underbrace{1.1 \cdots 1}_{n-1} \cdot \alpha + \underbrace{1.1 \cdots 1}_{n-1} \cdot \alpha + \cdots.$$

The terms not written down and indicated by $\cdots$ are all positive.

74

Thus the region from 1 to 1.1 is still too large for the construction of the graph of $y = x^{100}$. It is only in a region of length 0.01–0.02 that the values of the function do not differ from each other too greatly.

Let us now take the scale on the x-axis 100 times larger than on the y-axis. The graph of $y = x^{100}$ then is expanded in horizontal direction 100 times and will have the shape represented in Fig. 17a.

If n is still larger, it will be necessary to choose the region for accurate construction of the graph still smaller. In Fig. 17b, the reader can see the graph of $y = x^{1000}$ stretched 1000 times along the x-axis. It is remarkable that two identical curves were obtained!*

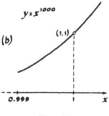

Fig. 17

EXERCISES

1. Construct the graph of the function $y = x^2 - x^4$ by two methods:

(a) subtraction of the graphs of $y = x^2$ and $y = x^4$;

(b) factoring the polynomial $x^2 - x^4$.

2. (a) There are given two increasing sequences

$$a_n: 0.001; 0.004; 0.009; \cdots,$$
$$b_n: 100; 300; 500; \cdots.$$

Can the first sequence catch up with the second (that is, can the inequality $a_n > b_n$ be satisfied for some n)?

(b) Answer the same question for the sequences

$$a_n: 0.001; 0.008; 0.027; \cdots,$$
$$b_n: 100; 400; 900; \cdots.$$

3. Determine how many solutions there are to the following equations:

(a) $x^3 = x^2 + 1$;   (b) $x^3 = x + 1$;

(c) $x^3 + 0.1 = 10x$;   (d) $x^5 - x - 1 = 0$.

4. (a) Draw graphs of the functions

$$y = x^3 - x; \qquad y = x^3 + x.$$

*Of course a difference between the curves exists; it is too small, however, to be observed in the graph.

75

(*b*) Select *a* and *b* so that in the graph of the function $y = ax^3 + bx$ there will be a dip 10 wide and not less than 100 deep.

6

While talking about the behavior of power functions $y = x^n$ for small values of the argument x, we noted that the graphs of these functions touch the x-axis at the origin. Let us dwell a little on this fact to clarify the exact meaning of the expression, a straight line touches a curve.

Actually, why do we say that the x-axis touches both the parabola $y = x^2$ and the cubic parabola $y = x^3$ (although the parabola $y = x^3$ is "pierced through" by the x-axis; Fig. 18)?

Why do we not consider the y-axis a tangent to the curve $y = x^2$, although it has only one point in common with it (see Fig. 18)?

Everything depends on what meaning is given to the expression "a straight line touches a curve," what is to be considered the basic defining property of a tangent, and what definition is to be given this concept. Up to now in the school geometry course we have become acquainted with only a single curve, the circle, and have learned only the definition of a tangent to a circle.

Let us try to understand in what respects the tangent to a circle differs from a secant. The following circumstances immediately become apparent: (1) a tangent has only one point in common with the circle while a secant has two (Fig. 19); (2) in the neighborhood of the point of tangency M the tangent line "lies nearer" the curve than any secant line through M. (Therefore, even if only part of the circle is drawn in the figure and we do not see whether in fact there is a second common point, we could still distinguish a tangent from a secant, Fig. 20).

Which of these two conditions is to be considered the more important and upon which should be based the definition of a tangent to any curve in general and not only to a circle?

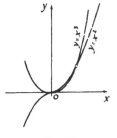
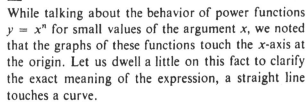

Fig. 18

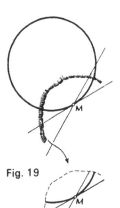

Fig. 19

Fig. 20

76

The first is simpler. It is possible to try to call a straight line having but one point in common with the curve a tangent.

For instance, through the vertex of the parabola $y = x^2$, beside the x-axis, there passes still another straight line having but one point in common with the parabola — this is the y-axis; the y-axis, however, is not called a tangent to the parabola $y = x^2$ (Fig. 18).

In Fig. 21 the situation is still worse. All straight lines lying within the angle AOB, intersect the curve only once! On the other hand, in Fig. 22 the straight line AB has two common points with the curve, yet, obviously has every right to be called a tangent. In fact, if having "clipped" the drawing, one looks at the region close to the point of contact O, the situation of the curve relative to the straight line will be of precisely the same character as the situation of the parabola relative to the x-axis (Fig. 23).

Consequently, it is reasonable to choose as a basic defining property of a tangent the fact that the tangent attaches itself closely to the curve.

Thus, for instance, it is natural to consider that the cubic parabola $y = x^3$ is tangent to the x-axis at the origin: for the parabola $y = x^3$ attaches itself closely to the x-axis at the origin (still more closely than the parabola $y = x^2$).

In order to give the definition of a tangent we must formulate what is meant by a straight line "closely attaching itself" to a curve. Let us examine again the parabola $y = x^2$. Between the x-axis (the straight line $y = 0$) and the parabola there cannot pass a single straight line: any straight line $y = kx$ (for $k > 0$) in some region runs above the parabola and intersects it once more.

If, rotating the straight line clockwise, we diminish k, the second point (the point M in Fig. 24) approaches the first (the point O) and finally coincides with it. At this moment the straight line turns from a secant into a tangent.

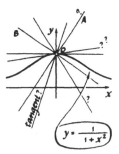

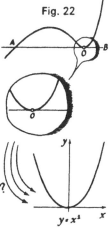

Fig. 21

Fig. 22

Fig. 23

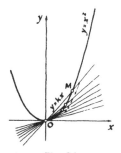

Fig. 24

77

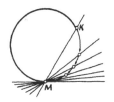

Fig. 25

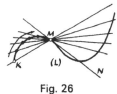

Fig. 26

Fig. 27

Fig. 28

The same can be seen in any case of tangency. In Fig. 25 the secant MK is drawn through the point M of the circle. If the point K is drawn nearer to M, the secant will turn about the point M and, finally when the point K coincides with the point M, will turn into a tangent to the circle at the point M. Then it will not have any other points in common with the circle. This circumstance, however, is nonessential and of secondary importance.

Thus we adopt the following definition of a tangent.

Definition. Suppose there is some curve (L) and a point M on this curve (Fig. 26). Let us draw a straight line MK through M and any other point K of the curve (L) (such a straight line, passing through any two points of the curve, is called a secant; it may also intersect the curve in some other points). If the point K is now moved along the curve (L) so as to approach the point M, the secant MK will turn about the point M. If finally, when the point K coincides with the point M, the straight line coincides with some definite straight line MN (Fig. 26), then this straight line MN is called a tangent to the curve (L) at the point M.

Thus the essential difference of a straight line tangent to some curve at the point M from other straight lines passing through the same point is that for the tangent its common point with the curve — the point of tangency — is a double point resulting from the merging of two approaching points of intersection.

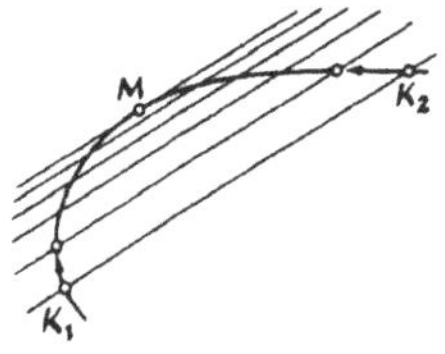

It is not necessary here that one of these points remain fixed: both points of intersection may move toward each other and coincide at the point of tangency (Fig. 27). Occasionally not two but three points merge in the point of tangency (Fig. 28).

Remark 1. In the definition nothing is stated about the number of common points of the curve and the tangent to it. This number can be arbitrary. In Fig. 29a the reader sees a straight line touching the curve (G) at the point M and intersecting it in two more points, and in Fig. 29b a straight line MN, which touches the curve at several points at a time.

78

Remark 2. In the definition of tangent it is assumed that the point K can approach the point M in any manner. In all cases the secant must tend to the same straight line, which is then called the tangent. If in different methods of approach of K to M the secant line tends to different straight lines the curve is said to have no tangent at this point.

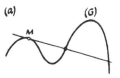

EXERCISES

1. Find the tangent at the point $O(0, 0)$ to the parabola $y = x^2 + x$.

Solution. Let us take some point M on the parabola with coordinates (a, b). Obviously, $b = a^2 + a$. Let us draw a straight line through the points O and M.

The equation of this straight line has the form $y = kx$. At $x = a$ we have $y = a^2 + a$, hence $k = a + 1$, and the equation of the secant is $y = (a + 1)x$. We shall now make the point $M(a, b)$ approach the point $O(0, 0)$. When the point M coincides with the point O, its abscissa a vanishes, and the secant $y = (1 + a)x$ becomes the tangent $y = x$.

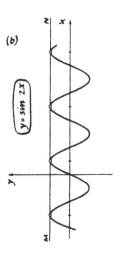

Fig. 29

Answer. The equation of the tangent is $y = x$.

2. Find the tangent to the parabola $y = x^2 + x$ at the point $A(1, 2)$. ⊕

3. Which of the straight lines parallel to the straight line $y = x$ is tangent to the parabola $y = -x^2 + 1$? ⊕

4. (a) Prove that the straight line $y = 0$ is the tangent to the curve

$$y = x^3 + x^2 \text{ at the origin.}$$

(b) Find the tangent at the point $O(0, 0)$ to the curve

$$y = x^3 - 2x.$$

5. For what cubic polynomials $y = ax^3 + bx^2 + cx$ does the x-axis serve as tangent at the origin?

CHAPTER 7

Rational Functions

1

Rational functions are functions that can be represented in the form of a quotient of two polynomials.

Examples of rational functions are

$$y = \frac{x^3 - 5x + 3}{x^6 + 1}, \qquad y = \frac{(x - 1)^2(x + 1)}{x^2 + 3},$$

$$y = x^2 + 3 - \frac{1}{x - 1}.^*$$

The linear fractional function

$$y = \frac{ax + b}{cx + d},$$

analyzed in Chapter 5, is rational. It is the quotient

$^*y = x^2 + 3 - 1/(x - 1)$ is a rational function, since it can be written in the form of a ratio of two polynomials:

$$x^2 + 3 - \frac{1}{x - 1} = \frac{(x^2 + 3)(x - 1) - 1}{x - 1}.$$

of two linear functions — polynomials of the first degree.

If the function $y = f(x)$ is the quotient of two polynomials of a degree higher than the first, then its graph will, as a rule, be more complicated, and constructing it accurately with all its details will sometimes be difficult. Frequently, however, it is sufficient to use methods analogous to those with which we have already become familiar.

2

Let us analyze some examples. We construct the graph of the function

$$y = \frac{x - 1}{x^2 + 2x + 1}.$$

First, let us notice that at $x = -1$ the function is not defined (since the denominator of the fraction $x^2 + 2x + 1$ equals zero at $x = -1$). For x close to -1, the numerator of the fraction $x - 1$ is approximately equal to -2, while the denominator $(x + 1)^2$ is positive and small in absolute value. Hence, the whole fraction $(x - 1)/(x + 1)^2$ will be negative and large in absolute value (and the more so, the closer x is to the value $x = -1$).

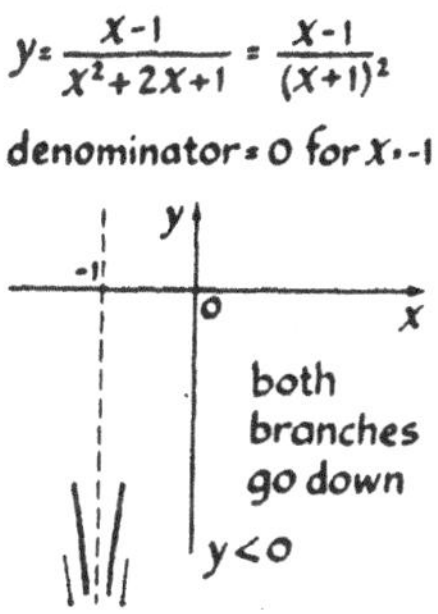

Conclusion. The graph splits into two branches (since there is no point on it with abscissa equal to -1); both branches go down as x approaches -1 (Fig. 1).

Let us consider the numerator. It vanishes at $x = 1$. Hence at the point $x = 1$ the graph intersects the x-axis. Having also drawn the point of intersection with the y-axis (at $x = 0, y = -1$), we can get an approximate idea of how the graph behaves in its central part (Fig. 2).

We still must find out what happens to the function at values of x large in absolute value.

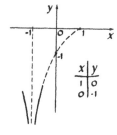

$y = \dfrac{x-1}{x^2+2x+1} = \dfrac{x-1}{(x+1)^2}$

denominator = 0 for x = -1

both branches go down

$y < 0$

Fig. 1

x	y
1	0
0	-1

Fig. 2

81

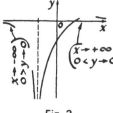

Fig. 3

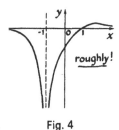

Fig. 4

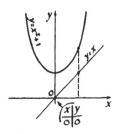

Fig. 5

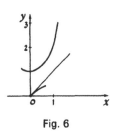

Fig. 6

If x is positive and increases, the numerator and denominator of the fraction increase. But since the numerator is of the first degree, while the denominator is of the second degree, the denominator, for large x, increases considerably faster than the numerator. Therefore, as x increases to infinity, the function $y = (x - 1)/(x^2 + 2x + 1)$ gets closer and closer to zero. In this way the right branch of the graph, to the right of the point $x = 1$, rises a little above the x-axis (Fig. 3), and then again starts to drop and will approach the x-axis.

Analogous considerations show that the left branch of the curve also approaches the x-axis as x increases in absolute value, except not from above but from below (Fig. 3). Later we shall show (see page 84) how to find the exact point where the right branch of the curve reaches its highest point.

From the above-mentioned details the general form of the graph can be found (Fig. 4).

3

Let us construct the graph of the function

$$y = \frac{x}{x^2 + 1}.$$

For convenience let us first draw the graphs of the numerator $y = x$ and the denominator $y = x^2 + 1$ (Fig. 5). For the construction of the graph of our function it is necessary to divide the values of the numerator by the values of the denominator.

At $x = 0$ the numerator is equal to zero — the graph passes through the origin. Let us go to the right (that is, we consider positive values of the argument). Since, for very small x, the value of x^2 is much smaller than x, as the graph leaves the origin, the denominator will for some time be almost equal to 1 (somewhat larger than 1); therefore the whole function will be approximately equal to the numerator x (slightly smaller than the numerator): the graph runs alongside the straight line $y = x$, gradually falling below it (Fig. 6).

Soon, however, $x^2 + 1$ starts to grow faster than x, the denominator leaves the numerator behind, and the fraction starts to decrease: the graph turns downward (Fig. 7).

Since the numerator is of the first degree, and the denominator contains an x^2-term, for large x the denominator grows faster than the numerator. Therefore, as x increases, the fraction becomes smaller and smaller — the graph approaches the x-axis (Fig. 8).

The left half of the graph can be obtained in an analogous way if it is observed that the given function is odd. The general form of the graph is shown in Fig. 9.

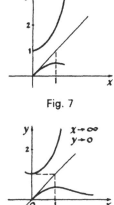

Fig. 7

Fig. 8

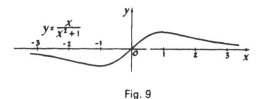

$y = \frac{x}{x^2+1}$

Fig. 9

4

Let us return to the graph of the function that we have constructed,

$$y = \frac{x}{x^2 + 1},$$

and analyze another interesting question, using this example. Let us try to find the highest point of the right half of the graph, exactly (and, hence, also the lowest point of the left half).

Obviously, our curve cannot rise very high, because the denominator $x^2 + 1$ starts quite rapidly to outgrow the numerator x. Let us find out whether the curve can reach a height equal to 1, that is, whether for some x, the value of y can be equal to 1.

Since $y = x/(x^2 + 1)$, it is necessary to solve the equation $x/(x^2 + 1)$ or the equation $x^2 - x + 1 = 0$. This equation does not have any real roots. (Check it!) Hence, on the graph there are no points with the ordi-

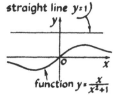

straight line $y=1$)

function $y = \frac{x}{x^2+1}$

Fig. 10

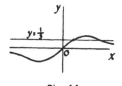

$y = \frac{1}{3}$

Fig. 11

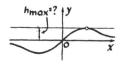

$h_{max}=?$

Fig. 12

nate $y = 1$ — the graph does not intersect the straight line $y = 1$ (Fig. 10).

Let us find out whether the curve reaches a height equal to $\frac{1}{3}$. For this purpose it is necessary that $x/(x^2 + 1) = \frac{1}{3}$, or $x^2 - 3x + 1 = 0$. This equation has two real roots (check it!), and therefore our graph has two points with ordinates equal to $\frac{1}{3}$; that is, it intersects the straight line $y = \frac{1}{3}$ at two points (Fig. 11).

In order to find the highest point, one must know for which largest h the equation $x/(x^2 + 1) = h$ will have a solution (Fig. 12).

Let us replace $x/(x^2 + 1) = h$ by the quadratic equation

$$hx^2 - x + h = 0.$$

This equation has a solution when

$$1 - 4h^2 \geq 0.$$

Hence, we can find the greatest height to which our graph can rise:

$$h = \tfrac{1}{2}.$$

Let us find at which x this largest value of y is obtained. Since $y = x/(x^2 + 1)$, $x/(x^2 + 1) = \frac{1}{2}$, $x^2 - 2x + 1 = 0$,* hence $x = 1$.

Thus the highest point of the graph is the point $(1, \tfrac{1}{2})$.

EXERCISE

Find the largest ordinate of the graph of $y = (x - 1)/(x + 1)^2$ (see Chapter 2).

5

Let us construct the graph of the function

$$y = \frac{x^2 + 1}{x}.$$

*Was it mere coincidence that a complete square was obtained?

Its general form can easily be drawn if it is noticed that

$$\frac{x^2 + 1}{x} = \frac{1}{x/(x^2 + 1)},$$

and, consequently, we have arrived at a problem that we have already solved: constructing the graph of $y = 1/f(x)$ from the graph of $y = f(x)$ (see pp. 56 and 57).

We get approximately the picture in Fig. 13.

Let us construct the graph by a different method. In doing so, we shall be able to explain another interesting peculiarity of this curve.

Let us divide the numerator by the denominator:

$$\frac{x^2 + 1}{x} = x + \frac{1}{x}.$$

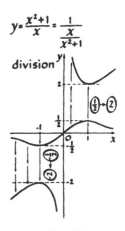

$$y = \frac{x^2+1}{x} = \frac{1}{\frac{x}{x^2+1}}$$

division

Fig. 13

Now we shall construct the graph of $y = x + (1/x)$ by "addition" of the known graphs of $y = x$ and $y = 1/x$ (Fig. 14).

We saw that the graph of $y = (x^2 + 1)/x$ has the y-axis as a vertical asymptote, which the graph approaches when x decreases in absolute value. It is now obvious that this graph also has an inclined asymptote, the straight line $y = x$ (this straight line is approached when x increases without bound).

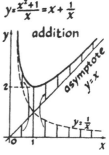

$$y = \frac{x^2+1}{x} = x + \frac{1}{x}$$

addition

asymptote $y = x$

$y = \frac{1}{x}$

Fig. 14

EXERCISES

1. Check that the graph of $y = (x^2 + 1)/x$ is symmetric with respect to the origin.

2. Find the coordinates of the lowest point of the right-hand branch of the graph of $y = (x^2 + 1)/x$.

The answer to the second exercise is clear from the first method by which this graph was constructed (see Fig. 13): The lowest point of the graph of $y = (x^2 + 1)/x$ is obtained for the x for which the graph of $y = x/(x^2 + 1)$ reaches its highest point, that is, for $x = 1$. The smallest ordinate value of the graph of $y = (x^2 + 1)/x$ is thus equal to 2.

We have obtained an interesting inequality: *for positive x* (the right-hand part of the graph was considered) we always have

$$x + \frac{1}{x} \geq 2.$$

Problems

1. Prove the inequality

$$x + \frac{1}{x} \geq 2 \text{ for } x > 0 \tag{1}$$

directly.

2. Prove the inequality

$$\frac{a + b}{2} \geq \sqrt{ab}. \tag{2}$$

It expresses this fact: "The arithmetic mean of two positive numbers a and b is always greater than or equal to the geometric mean of these numbers."

Inequality 1 is a particular case of Inequality 2. For what a and b?

3. The inequality $x + (1/x) \geq 2$ is used in solving the well-known problem of the "honest merchant." An honest merchant knew that the scales on which he was weighing his merchandise were inaccurate, because one beam was somewhat longer than the other (at that time they were still using scales as shown in Fig. 15). What was he to do? Cheating his customers would be dishonest, but he did not want to hurt himself either. The merchant decided that he would weigh out half the merchandise to each buyer on one scale and the second half on the other scale.

The question is: Did the merchant gain or lose, as a result?

Fig. 15

6

In the following example let us take the function

$$y = \frac{1}{x+1} + \frac{1}{x-1}.$$

It is already in the form of the sum of two functions, and of course it is possible to construct its graph by adding the graphs of $y = 1/(x + 1)$ and $y = 1/(x - 1)$. In this case, however, it is perhaps possible to find the general form of the graph from the following considerations:

(*a*) The function is not defined at $x = 1$ and $x = -1$, and therefore the curve separates into three branches (Fig. 16).

(*b*) As x approaches one, the second term, and hence the whole function as well, increases in absolute value, and therefore the branches of the graph move away from the x-axis, approaching the straight line $x = 1$. On the right of $x = 1$ the curve goes up, and on the left it goes down (Fig. 17).

An analogous picture results near the straight line $x = -1$:

(*c*) $y = 0$ at $x = 0$; the curve passes through the origin (Fig. 18).

(*d*) For numbers large in absolute value, both terms are small in absolute value, and both extreme branches of the graph approach the x-axis: the right from above, and the left from below (Fig. 19).

Combining all this information, we can obtain the general form of the graph (Fig. 20).

Show that this graph is symmetric with respect to the origin.

7

The examples discussed show that even for the construction of the same graph, different methods can be employed. Therefore, we shall now give a few more examples for the construction of graphs and

$y = \frac{1}{x+1} + \frac{1}{x-1}$

breaks at points
$x = -1$
$x = +1$

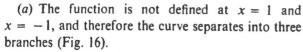

asymptotes

Fig. 16

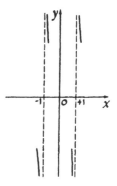

Fig. 17

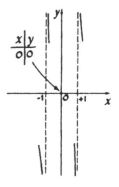

x	y
0	0

Fig. 18

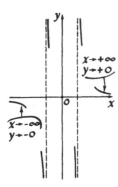

Fig. 19

shall not deprive the reader of the pleasure of selecting for himself the most suitable methods for constructing them.

EXERCISES

1. Construct the graph of

$$y = \frac{1}{x} + \frac{1}{x+1} + \frac{1}{x+2}.$$

Into how many pieces does the curve split?

2. (*a*) Construct the graph of

$$y = \frac{1}{x-1} - \frac{1}{x+1}.$$

(*b*) Construct the graph of

$$y = \frac{1}{x} - \frac{1}{x+2}.$$

Indicate the axis of symmetry of this curve.

3. Construct the graphs of

(*a*) $y = \dfrac{1}{(x-1)(x-2)}$;

(*b*) $y = \dfrac{1}{(x-1)(x-2)(x+1)}$.

(*c*) $v = x + \dfrac{1}{x^2}$.

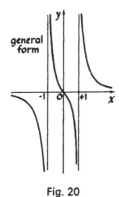

general form

Fig. 20

88

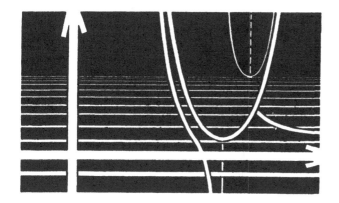

Problems for Independent Solution

1. Construct the graphs of the functions:

(a) $y = x(1 - x) - 2$;

(b) $y = x(1 - x)(x - 2)$;

(c) $y = \dfrac{4 - x}{x^3 - 4x}$;

(d) $y = \dfrac{2|x| - 3}{3|x| - 2}$;

(e) $y = \dfrac{1}{4x^2 - 8x - 5}$;

(f) $y = \dfrac{1}{x^3 - 5x}$;

(g) $y = \dfrac{1}{x^2} + \dfrac{1}{x - 1}$;

(h) $y = \dfrac{1}{x^2} + \dfrac{1}{x^3}$; $\oplus$

(i) $y = (2x^2 + x - 1)^2$;

(j) $y = |x| + \dfrac{1}{1 + x^2}$;

(k) $y = \dfrac{x^2 + 2x}{x^2 + 4x + 3}$;

(l) $y = \dfrac{x^2 - 2x + 4}{x^2 + x - 2}$;

(m) $y = (x - 3)|x + 1|$;

(n) $y = |x - 2| + 2|x| + |x + 2|$;

(o) $y = \left[\dfrac{1}{x} \right]$;

(p) $y = \dfrac{|x + 1| - x}{|x - 2| + 3}$;

(q) $y = \dfrac{x}{[x]}$ $\oplus$

(r) Construct graphs of the linear fractional functions of the form

$$y = \frac{3x + a}{2x + 2}$$

for different values of a.

2. The function $y = f(x)$ is defined by the following rule:

$$f(x) = \begin{cases} 1 \text{ for } x > 0, \\ 0 \text{ for } x = 0, \\ -1 \text{ for } x < 0. \end{cases}$$

This function is frequently encountered, and therefore there is a special notation for it:

$$y = \text{sign } x \text{ (read "signum } x\text{").}$$

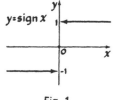

y=sign x

Fig. 1

The graph of this function is represented in Fig. 1. For $x \neq 0$ the function sign x can be defined by the formula $y = x/|x|$.*

Draw graphs of the functions:

$$y = \text{sign}^2 x; \quad y = (x - 1) \text{ sign } x; \quad y = x^2 \text{ sign } x.$$

3. The general form of the graph of the function that is the quotient obtained by dividing one quadratic

*Why only for $x \neq 0$?

90

trinomial by another,

$$y = \frac{ax^2 + bx + c}{x^2 + px + q},$$

depends on how many and what roots the numerator and denominator have.

(a) Construct the graphs of the functions:

$$y = \frac{4x^2 - 8x + 3}{x - x^2},$$

$$y = \frac{x^2 - 2x + 1}{x^2 + 2}, \qquad y = \frac{3x^2 - 10x + 3}{x^2 - x - 6}.$$

(b) What is the form of the graph of the function

$$y = \frac{ax^2 + bx + c}{x^2 + px + q},$$

if both roots of the numerator are less than the roots of the denominator? ⊕

(c) Analyze all possible cases and draw the possible types of graphs for functions of the form

$$y = \frac{ax^2 + bx + c}{x^2 + px + q}.$$

Try not to omit a single case, and give one example for each type.

4. Construct the graph of the function $y = \sqrt{3}\, x$.

(a) Prove that it cannot pass through any point whose coordinates are integers, except the point $(0, 0)$.

If one square is taken as scale unit, then the vertices of the squares will be these "integral" points. Take the origin close to the lower left corner of a notebook and draw the straight line $y = \sqrt{3}\, x$ accurately. (At what angle with the x-axis must it be drawn?) Some of the integral points turn out to be very close to this straight line. Use this to find approximate values for $\sqrt{3}$ in the form of ordinary fractions. Compare the values obtained with the tabular one: $\sqrt{3} \approx 1.7321$.

(b) A hard problem: Prove that there is an integral point at a distance of less than $\frac{1}{1000}$* from the straight line $y = \sqrt{3}\, x$.

In solving Problems 5 to 9 make use of the graphs of suitably chosen functions.

5. How many solutions are there for the following equations?

(a) $-x^2 + x - 1 = |x|$;

(b) $|3x^2 + 12x + 9| + x = 0$;

(c) $\dfrac{1}{x^2 - x + 1} = x$;

(d) $|x - 1| + |x - 2| + |x + 1| + |x + 2| = 6$;

(e) $x(x + 1)(x + 2) = 0.01$;

(f) $|x + 3| = |x + 2|(x^2 - 1)$;

(g) $[x] = x$ in the interval $|x| < 3$;

(h) $\dfrac{1}{x} + \dfrac{1}{x + 1} + \dfrac{1}{x + 2} = 100$.

6. Solve the equations:

(a) $2x^2 - x - 1 = |x|$;

(b) $|2x^2 - x - 1| - x = 0$;

(c) $|x| = |x - 1| + |x - 2|$.

7. (a) Determine how many solutions the equation

$$|1 - |x|| = a$$

can have for different values of a.

(b) The same question for the equation

$$x^2 + \frac{1}{x} = a.†$$

*The number $\frac{1}{1000}$ can be replaced by any other number. Then it will be proved that however small a number is taken, there is a point with integral coordinates removed at a distance from the straight line $y = \sqrt{3}\, x$ that is less than this number.

†The value of a separating the different cases can be found approximately from the graph.

8. Solve the inequalities:

(a) $\dfrac{2 - x}{x^2 + 6x + 5} > 0$;

(b) $x \leq |x^2 - x|$;

(c) $|x| + 2|x + 1| > 3$.

9. Find the largest value of the function and determine for which values of x it is reached:

(a) $y = x(a - x)$;

(b) $y = |x|(a - |x|)$;

(c) $y = x^2(a - x^2)$;

(d) $y = \dfrac{x^2 + 4}{x^2 + x + 1}$;

(e) $y = 1 - \sqrt{2x}$ in the interval $|x| \leq \sqrt{2}$;

(f) $y = -x^2 + 2x - 2$ in the interval $-5 \leq x \leq 0$;

(g) $y = \dfrac{x + 3}{x - 1}$ in the interval $x \geq 2$.

10. Two roads intersect at a right angle. Two cars drive toward the intersection: on the first road at a speed of 60 km/hr, on the second at a speed of 30 km/hr. At noon both cars are 10 km from the intersection.

At what moment will the distance between the cars be least? Where will the cars be at this moment?

11. Among all right triangles with given perimeter p, find the triangle having the largest area.

12. Suppose $y = f(x)$ is an even function, and $y = g(x)$ is an odd function. What can be said about the parity of the following functions?

$y = f(x) + g(x)$; $\quad y = f(x)g(x)$;

$y = |g(x)|$; $\quad y = f(x) - g(x)$;

$y = f(|x|) - g(x)$; $\quad y = f(x) - g(|x|)$.

13. Find all even and all odd functions of the form:

(a) $y = kx + b$;

(b) $y = \dfrac{px + q}{x + r}$;

(c) $y = \dfrac{ax^2 + bx + c}{x^2 + px + q}$.

14. The function $y = x^4 - x$ is neither even nor odd. However, this function is easily represented in the form of a sum of an even function $y = x^4$ and of and odd function $y = -x$ (Fig. 2).

(a) Represent the function $y = 1/(x^4 - x)$ as a sum of an even and an odd function.

(b) Prove that any function $f(x)$ can be represented as a sum of an even and an odd function. ⊕

15. Through any two points with different abscissas there passes a straight line (the graph of a linear function $y = kx + b$). Analogously, through any three points with different abscissas and which are not on one straight line, it is possible to draw a parabola, the graph of a function $y = ax^2 + bx + c$.

Find the coefficients of the quadratic trinomial $ax^2 + bx + c$, whose graph passes through the points:

(a) $(-1, 0)$; $(0, 2)$; $(1, 0)$;

(b) $(1, 0)$; $(4, 0)$; $(5, 6)$;

(c) $(-6, 7)$; $(-4, -1)$; $(-2, 7)$;

(d) $(0, -4)$; $(1, -3)$; $(2, -1)$;

(e) $(-1, 9)$; $(3, 1)$; $(6, 16)$.

16. (a) Carry out a similarity transformation of the parabola $y = x^2$, choosing the center of similarity at the origin and a ratio of similarity equal to 2. What curve is obtained?

(b) What similarity transformation transforms the curve $y = x^2$ into the curve $y = 5x^2$?

(c) Using the result of Problem 16b, find the focus and the directrix of the parabola $y = 4x^2$. (Definitions of directrix and focus are on p. 43.)

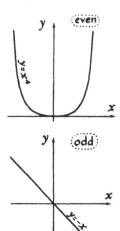

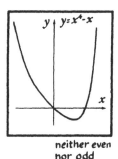

neither even
nor odd

Fig. 2

94

(d) Prove that all parabolas $y = ax^2 + bx + c$ are geometrically similar.

17. Prove that the point $F(0, \frac{1}{4})$ is the focus and the straight line $y = -\frac{1}{4}$ is the directrix of the parabola $y = x^2$; that is, any point of this parabola is equidistant from the point $F(0, \frac{1}{4})$ and the straight line $y = -\frac{1}{4}$.

Hint. Take, on the parabola, some point M with the coordinates (a, a^2). Write down the distance of this point from the point $F(0, \frac{1}{4})$, using the formula for the distance between two points.* Then write down the distance from the point $M(a, a^2)$ to the straight line $y = -\frac{1}{4}$.†

Prove the equality of the two resulting expressions.

18. Prove that the points $F_1(\sqrt{2}, \sqrt{2})$ and $F_2(-\sqrt{2}, -\sqrt{2})$ are the foci of the hyperbola $y = 1/x$; that is, the difference of the distances from any point of this hyperbola to the points F_1 and F_2 is constant in absolute value.

Hint. Take an arbitrary point $M(a, 1/a)$ on the hyperbola $y = (1/x)$. Express the distances of this point from the point $F_1(\sqrt{2}, \sqrt{2})$ and from the point $F_2(-\sqrt{2}, -\sqrt{2})$ in terms of a. Show that the absolute value of this difference is the same for all values of a (and, hence, does not depend on the choice of the point on the hyperbola).

19. On pages 96 and 97, seventeen graphs and as many formulas are given. The problem is to determine which formula belongs to which of the numbered graphs. Among these graphs the reader can find the answers to exercises.

*The distance between the two points $A(x_1, y_1)$ and $B(x_2, y_2)$ is given by the formula $d(A, B) = \sqrt{(x_1 - x_2)^2 + (y_1 - y_2)^2}$.
†The distance from the point $A(x_1, y_1)$ to the straight line $y = c$ equals $|y_1 - c|$.

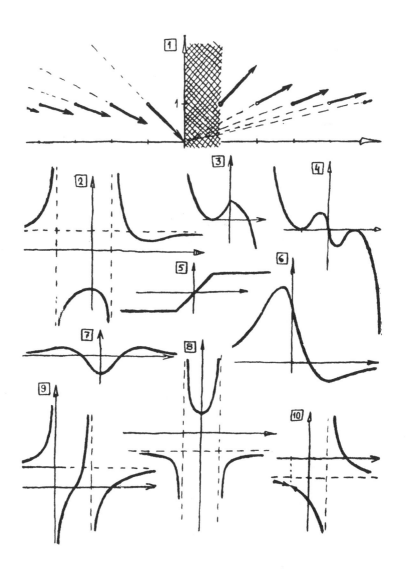

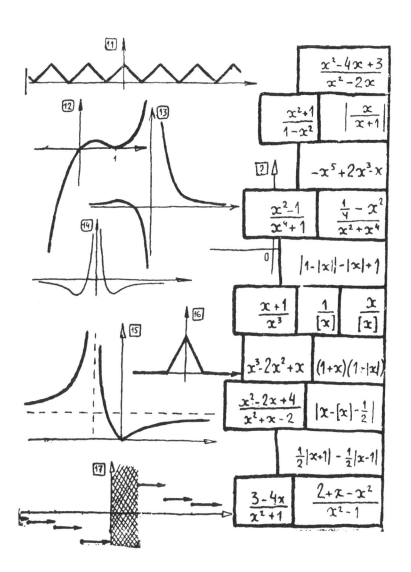

$$\boxed{11}$$

$$\frac{x^2-4x+3}{x^2-2x}$$

$$\frac{x^2+1}{1-x^2} \qquad \left|\frac{x}{x+1}\right|$$

$$\boxed{12} \qquad \boxed{13}$$

$$\boxed{2} \qquad -x^5+2x^3-x$$

$$\frac{x^2-1}{x^4+1} \qquad \frac{\frac{1}{4}-x^2}{x^2+x^4}$$

$$\boxed{14}$$

$$0 \qquad \left|1-|x|\right|-|x|+1$$

$$\frac{x+1}{x^3} \qquad \frac{1}{[x]} \qquad \frac{x}{[x]}$$

$$\boxed{15} \qquad \boxed{16}$$

$$x^3-2x^2+x \qquad (1+x)(1-|x|)$$

$$\frac{x^2-2x+4}{x^2+x-2} \qquad \left|x-[x]-\frac{1}{2}\right|$$

$$\frac{1}{2}|x+1|-\frac{1}{2}|x-1|$$

$$\boxed{17}$$

$$\frac{3-4x}{x^2+1} \qquad \frac{2+x-x^2}{x^2-1}$$

20. Figure 3 represents the graph of the function $y = f(x)$.

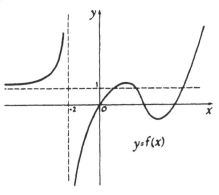

Fig. 3

Sketch the graphs of the following functions:

(a) $y = f(x) - 2$; (b) $y = f(x + 2)$;

(c) $y = |f(x)|$; (d) $y = f(|x|)$;

(e) $y = -3f(x)$; (f) $y = \dfrac{1}{f(x)}$;

(g) $y = (f(x))^2$; (h) $y = f(-x)$;

(i) $y = x + f(x)$; (j) $y = \dfrac{f(x)}{x}$.

21. A square with side a is drawn in the plane (Fig. 4). The curve L_s is the locus of all points the least distance of which from some point of the square is equal to S. Let us denote the area bounded by the curve L_s by $P(S)$.

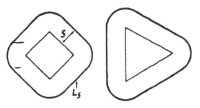

Fig. 4

98

(a) Find $P(S)$ as a function of S.

(b) Solve a problem analogous to Problem 21a, but instead of a square, take a rectangle with sides a and b.

(c) The same problem for the triangle with sides a, b, and c.

(d) The same problem for the circle of radius r.

22. Can you find a rule for the resulting expressions for $P(S)$? Write down a general formula for any convex figure. Does this formula hold for non-convex figures?

23. We shall examine quadratic equations of the form $x^2 + px + q = 0$; each such equation is completely determined by the two numbers p and q. Let us agree to represent this equation by the point in the plane with coordinates (p, q). For example, the equation $x^2 - 2x + 3 = 0$ is represented by the point $A(-2, 3)$; the equation $x^2 - 1 = 0$ by the point $B(0, -1)$.

(a) What equation corresponds to the origin?

(b) Draw the set of points corresponding to those equations whose roots have a sum equal to zero.

(c) Randomly select a point in the plane. If the equation corresponding to this point has two real roots, mark the point with a green pencil. If the equation does not have any real roots, mark the point with a red pencil. Take a few more points and do the same with them. Can you state which part of the plane is occupied by "green" points and which by "red" ones? What line separates the "green" points from the "red" ones? How many roots have the equations corresponding to the points of this line?

(d) What point set corresponds to those equations whose roots are real and positive?

(e) By what point can an equation be represented if it is known that one of its roots is equal to 1?

24. Up to some moment a car was traveling in uniformly accelerated motion and then started to travel in uniform motion (at the speed attained by it).

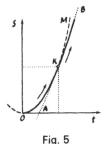

Fig. 5

The graph of the motion of this car is depicted in Fig. 5. Prove that the straight line AB is a tangent to the parabola OKM.

25. (a) Using graphs determine the number of solutions of the following cubic equations:

(1) $0.01x^3 = x^2 - 1$,

(2) $0.001x^3 = x^2 - 3x + 2$.

(b) Find approximate values of the roots of these equations.

26. (a) On p. 34 there is a diagram showing that the graph of the polynomial $y = x^4 - \frac{5}{2}x^2 + \frac{9}{16}$ is obtained from the graph of the polynomial $y = x^4 - 2x^3 - x^2 + 2x$ by translation along the x-axis. Find the value of this translation.

(b) Solve the equation of the fourth degree

$$x^4 - 6x^3 + 7x^2 + 6x - 8 = 0.$$

Hint. Translate the graph of the polynomial $x^4 - 6x^3 + 7x^2 + 6x$ along the x-axis so that it becomes the graph of some biquadratic polynomial.

(c) Under what conditions does the curve

$$y = x^4 + bx^3 + cx^2 + d$$

have an axis of symmetry? ⊕

27. Let us solve Problem 4 on page 38. There are $19 + 9 + 26 + 8 + 18 + 11 + 14 = 105$ matches in all. Therefore it is necessary to obtain a distribution of $105 \div 7 = 15$ matches in each box.

Let us denote by x the number of matches that must be shifted from the first box to the second. (It may, of course, be necessary to shift matches from the second box to the first, in which case x will be negative.) After we have shifted x matches from the first box to the second, there will be $x + 9$ matches in the second box.

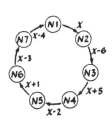

Fig. 6

100

Therefore it is necessary to move $x - 6$ matches from the second to the third, $x + 5$ matches from the third to the fourth. Similarly, $x - 2$ are shifted from the fourth box to the fifth, $x + 1$ from the fifth to the sixth, $x - 3$ from the sixth to the seventh, and finally $x - 4$ matches from the seventh to the first (Fig. 6).

Let us now denote by S the total number of shifted matches:

$$S = |x| + |x - 6| + |x + 5| + |x - 2|$$
$$+ |x + 1| + |x - 3| + |x - 4|.$$

In this formula absolute-value signs were used because we are interested only in the number of matches transposed and not in the direction in which they were shifted.

We must now choose x so that S has the least value. Here the graph of the function $S = f(x)$ can be helpful (Fig. 7): The lowest point of the graph is the vertex A_4; that is, the function $S = f(x)$ assumes its least value at $x = 2$. Thus x has been found, and we can say how many matches must be moved, and where they will be moved (Fig. 8). In this manner the problem can also be solved, of course, for an arbitrary number (n) of boxes. For this purpose it is necessary, as in our example, to write down an expression for S. It will be of the form:

$$S = |x| + |x - a_1| + |x - a_2| + \cdots$$
$$+ |x - a_{n-1}|.$$

In order to find the required value of x in the case of an odd number n, the following simple rule can be used: The numbers $0, a_1, a_2, \cdots, a_{n-1}$ must be written down in increasing order, after which x is chosen equal to the number exactly in the center of this sequence of numbers (if n is odd, such a number

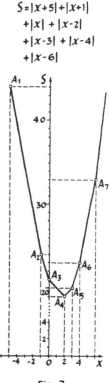

$$S = |x+5| + |x+1|$$
$$+ |x| + |x-2|$$
$$+ |x-3| + |x-4|$$
$$+ |x-6|$$

Fig. 7

Answer

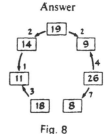

Fig. 8

can always be found). Consider how the graph looks when *n* is even.

Then try to state a rule for finding *x* in this case.

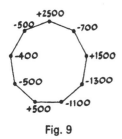

Fig. 9

This problem, which is similar to a game, is connected with the practical problem of transportation along circular routes. Imagine a circular railroad with evenly spaced stations. At some stations there are coal storages; at others there are users of coal who must be furnished with all of this coal. Figure 9 indicates the stocks of coal at the storages and (with a minus sign) the respective needs of the users.

Using the solution of the preceding problem, set up the most economical transportation plan.

Answers and Hints to Problems and Exercises Marked with the Sign ⊕

Exercise 2*b*, p. 17. Look for the answer among the graphs on pp. 96 and 97.

Exercise 3*b*, on p. 32, for the graph in Fig. 12. Look for the answer among the graphs on pp. 96 and 97.

Exercise 3, p. 37. Hint. This function assumes its minimum on an entire segment.

Problem 4 on p. 38. Solution in Problem 27 on p. 100.

Exercise 2*b* on p. 60. No, it does not. The rigorous proof of this fact is not very simple, and we shall not give it here. It is clear, however, that since the *x*-axis and the *y*-axis are asymptotes of this curve, the only possible axis of symmetry is the straight line $y = x$. It is easy to verify that this straight line is not an axis of symmetry.

Exercise 2 on p. 79. The equation of the tangent is $y = 3x - 1$.

Exercise 3 on p. 79. Hint. The system

$$y = x + a, y = -x^2 - 1$$

must have two coincident solutions.

Problem 1*h*, p. 89 and 1*q*, p. 90. Look for the answer among the graphs on pp. 96 and 97.

Problem 3*b*, p. 91.

Let us take a numerical example. Suppose the roots of the numerator are -5 and 0, and those of the denominator are $+2$ and $+4$. Then our function is of the form $y = [ax(x + 5)]/[(x - 2)(x - 4)]$. Let us take some concrete value of *a*, e.g., $a = 2$. The function

$$y = \frac{[2x(x + 5)]}{[(x - 2)(x - 4)]}$$

is not defined at $x = 2$ and $x = 4$. As *x* approaches these values, the denominator decreases, approaching zero; therefore, the function increases without bound in absolute value — the straight lines $x = 2$ and $x = 4$ are vertical asymptotes of the graph.

The function is equal to zero at $x = 0$ and $x = -5$. Let us mark two points of the graph on the *x*-axis: $(0, 0)$ and $(-5, 0)$.

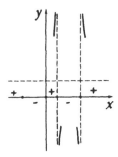

Fig. 10

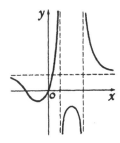

Fig. 11

The four "special" values of the argument, $x = -5$, 0, 2, 4, divide the x-axis into 5 intervals. As x passes the boundary of any interval, the function changes its sign (vanishing or "going off to infinity") (Fig. 10).

We must explain the behavior of the function when the argument increases without bound in absolute value. Let us try to substitute large numbers for x (e.g., $x = 10,000$, $x = 1,000,000$, etc.). Since $2x^2$ will be considerably larger than $10x$, and x^2 considerably larger than $-6x + 8$, the fraction

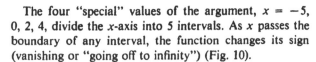

$$\frac{2x(x + 5)}{(x - 2)(x - 4)} = \frac{2x^2 + 10x}{x^2 - 6x + 8}$$

will be approximately equal to the ratio of the highest terms of numerator and denominator,

$$y = \frac{2x^2 + 10x}{x^2 - 6x + 8} \approx \frac{2x^2}{x^2} = 2,$$

and will be the closer to 2, the larger $|x|$ is. Hence, as x moves away from the origin, the graph approaches the horizontal straight line $y = 2$.

The general form of the graph is given in Fig. 11. In all cases when both roots of the denominator are larger than the roots of the numerator, the graph will be approximately of this form.

Problem 14, p. 94. As is often the case in mathematics, the problem is easier to solve in a general form than for a specified concrete function. Therefore, we shall first solve Problem b, and obtain the solution of a as a particular case.

Thus, suppose we are given some function $f(x)$. Let us suppose the problem is solved; that is, $f(x)$ is represented as a sum of an even function $g(x)$ and an odd function $h(x)$:

$$f(x) = g(x) + h(x). \tag{*}$$

Since this equality holds for all values of x, $-x$ can be substituted for x, and we obtain

$$f(-x) = g(-x) + h(-x). \tag{**}$$

Since the function $g(x)$ is even, and $h(x)$ is odd, $g(-x) = g(x)$ and $h(-x) = -h(x)$. Using this, we first add

104

Eqs. * and **, and then subtract one from the other; we find

$$f(x) + f(-x) = 2g(x), \quad f(x) - f(-x) = 2h(x).$$

From this the functions $g(x)$ and $h(x)$ can be found, and the desired decomposition of the function $f(x)$ into a sum of an even and an odd function can be obtained:

$$f(x) = \frac{f(x) + f(-x)}{2} + \frac{f(x) - f(-x)}{2}. \quad (1)$$

Notice that the formal proof of the result we have obtained is even simpler: Write down the decomposition of Eq. 1, and check that it is identically satisfied for all x and that the first term on the right-hand side is an even function while the second is odd.

The solution of Problem a is obtained directly by Formula 1:

$$\frac{1}{x^4 - x} = \frac{x^2}{x^6 - 1} + \frac{1}{x^7 - x}.$$

Remark. If the function $y = f(x)$ is not defined for some values of x, then also the functions $g(x)$ and $h(x)$ will not be defined for all values of x. In this case, it may turn out that for some x the function $f(x)$ is defined, while $g(x)$ and $h(x)$ are not defined.

Exercise 26c, p. 100. $4d = b^3 + 2bc$.